Plumbing
Level Three

Trainee Guide
Third Edition

Upper Saddle River, New Jersey
Columbus, Ohio

National Center for Construction Education and Research

President: Don Whyte
Director of Curriculum Revision and Development: Daniele Stacey
Plumbing Project Manager: Daniele Stacey
Production Manager: Jessica Martin
Production Maintenance Supervisor: Debie Ness
Editors: Brendan Coote and Bethany Harvey
Desktop Publishers: Debie Ness and Jessica Martin

NCCER would like to acknowledge the contract service provider for this curriculum:
EEI Communications, Alexandria, Virginia.

This information is general in nature and intended for training purposes only. Actual performance of activities described in this manual requires compliance with all applicable operating, service, maintenance, and safety procedures under the direction of qualified personnel. References in this manual to patented or proprietary devices do not constitute a recommendation of their use.

Copyright © 2006, 2002, 1993 by the National Center for Construction Education and Research (NCCER), Gainesville, FL 32614-1104, and published by Pearson Education, Inc., Upper Saddle River, NJ 07458. All rights reserved. Printed in the United States of America. This publication is protected by Copyright and permission should be obtained from the NCCER prior to any prohibited reproduction, storage in a retrieval system, or transmission in any form or by any means, electronic, mechanical, photocopying, recording, or likewise. For information regarding permission(s), write to: NCCER Product Development, P.O. Box 141104, Gainesville, FL 32614-1104.

10 9 8 7
ISBN 0-13-227301-2

Preface

TO THE TRAINEE

Most people are familiar with plumbers who come to their home to unclog a drain or install an appliance. In addition to these activities, however, plumbers install, maintain, and repair many different types of pipe systems. For example, some systems move water to a municipal water treatment plant and then to residential, commercial, and public buildings. Other systems dispose of waste, provide gas to stoves and furnaces, or supply air conditioning. Pipe systems in power plants carry the steam that powers huge turbines. Pipes also are used in manufacturing plants, such as wineries, to move material through production processes.

Plumbers and their associated trades constitute one of the largest construction occupations, holding about 550,000 jobs. Theirs is also among the highest paid construction occupations. Even better, job opportunities are expected to be excellent as demand for skilled craftspeople is expected to outpace the supply of trained plumbers.[1]

We wish you success as you embark on your third year of training in the plumbing craft and hope that you'll continue your training beyond this textbook. As most of the half a million craftspeople employed in this trade can tell you, there are many opportunities awaiting those with the skills and the desire to move forward in the construction industry.

NEW WITH *PLUMBING LEVEL THREE*

NCCER and Prentice Hall are pleased to present the third edition of Plumbing Level Three. This edition presents a new design. Check out the opening pages to each of the nine modules in this textbook to see people in the plumbing trade competing in the annual Associated Builders and Contractors (ABC) Craft Olympics. ABC is a national association representing 23,000 merit shop construction and construction-related firms in 79 chapters across the United States. The annual ABC Craft Olympics competition spotlights the nation's top craft professionals. As an integral part of ABC's annual convention, craft students from chapters and member firm training programs around the country participate in the National Craft Olympics. During this intense two-day event, young men and women compete in one of 13 craft categories, including plumbing. For more information on ABC and the Craft Olympics visit www.abc.org.

We invite you to visit the NCCER website at www.nccer.org for the latest releases, training information, newsletter, and much more. You can also reference the Contren® product catalog online at www.crafttraining.com. Your feedback is welcome. You may email your comments to curriculum@nccer.org or send general comments and inquiries to info@nccer.org.

CONTREN® LEARNING SERIES

The National Center for Construction Education and Research (NCCER) is a not-for-profit 501(c)(3) education foundation established in 1995 by the world's largest and most progressive construction companies and national construction associations. It was founded to address the severe workforce shortage facing the industry and to develop a standardized training process and curricula. Today, NCCER is supported by hundreds of leading construction and maintenance companies, manufacturers, and national associations. The Contren® Learning Series was developed by NCCER in partnership with Prentice Hall, the world's largest educational publisher.

Some features of NCCER's Contren® Learning Series are as follows:

- An industry-proven record of success
- Curricula developed by the industry for the industry
- National standardization, providing portability of learned job skills and educational credits
- Compliance with Apprenticeship, Training, Employer, and Labor Services (ATELS) requirements for related classroom training (CFR 29:29)
- Well-illustrated, up-to-date, and practical information

NCCER also maintains a National Registry that provides transcripts, certificates, and wallet cards to individuals who have successfully completed modules of NCCER's Contren® Learning Series. *Training programs must be delivered by an NCCER Accredited Training Sponsor in order to receive these credentials.*

---

[1] U.S. Department of Labor, Bureau of Labor Statistics. *Occupational Outlook Handbook, 2004–05 Edition.* www.bls.gov/oco

Special Features of This Book

In an effort to provide a comprehensive user-friendly training resource, we have incorporated many different features for you to use. Whether you are a visual or hands-on learner, this book will provide you with the proper tools to get started in the plumbing industry.

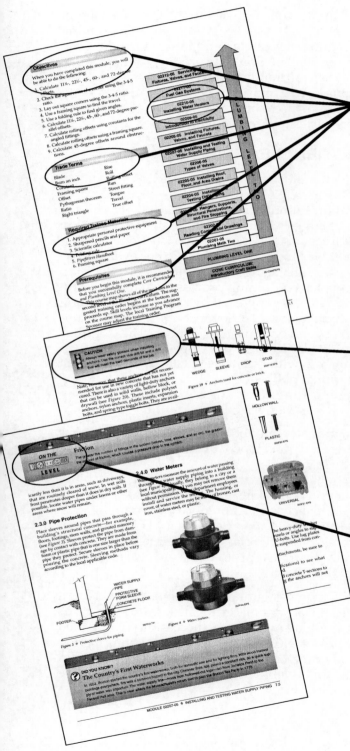

Introduction Page

This page is found at the beginning of each module and lists the Objectives, Trade Terms, Required Trainee Materials, Prerequisites, and Course Map for that module. The Objectives list the skills and knowledge you will need in order to complete the module successfully. The list of Trade Terms identifies important terms you will need to know by the end of the module. Required Trainee Materials list the materials and supplies needed for the module. The Prerequisites for the module are listed and illustrated in the Course Map. The Course Map also gives a visual overview of the entire course and a suggested learning sequence for you to follow.

Notes, Cautions, and Warnings

Safety features are set off from the main text in highlighted boxes and organized into three categories based on the potential danger of the issue being addressed. Notes simply provide additional information on the topic area. Cautions alert you of a danger that does not present potential injury but may cause damage to equipment. Warnings stress a potentially dangerous situation that may cause injury to you or a co-worker.

On the Level

The On the Level features offer technical hints and tips from the plumbing industry. These often include nice-to-know information that you will find helpful. On the Level also presents real-life scenarios similar to those you might encounter on the job site.

Did You Know?

The Did You Know? features introduce historical tidbits or modern information about the plumbing industry. Interesting and sometimes surprising facts about plumbing are also presented.

Illustrations and Photographs

Illustrations and photographs are used throughout each module to provide extra detail. These figures highlight important concepts from the text and provide clarity for complex instructions. Each figure is denoted in the text in *italic type* for easy reference.

Step-by-Step Instructions

Step-by-step instructions are used throughout to guide you through technical procedures and tasks from start to finish. These steps show you not only how to perform a task but also how to do it safely and efficiently.

Trade Terms

Each module presents a list of Trade Terms that are discussed within the text, defined in the Glossary at the end of the module. These terms are denoted in the text with **bold type** upon their first occurrence. To make searches for key information easier, a comprehensive Glossary of Trade Terms from all modules is found at the back of this book.

Review Questions

Review Questions are provided to reinforce the knowledge you have gained. This makes them a useful tool for measuring what you have learned.

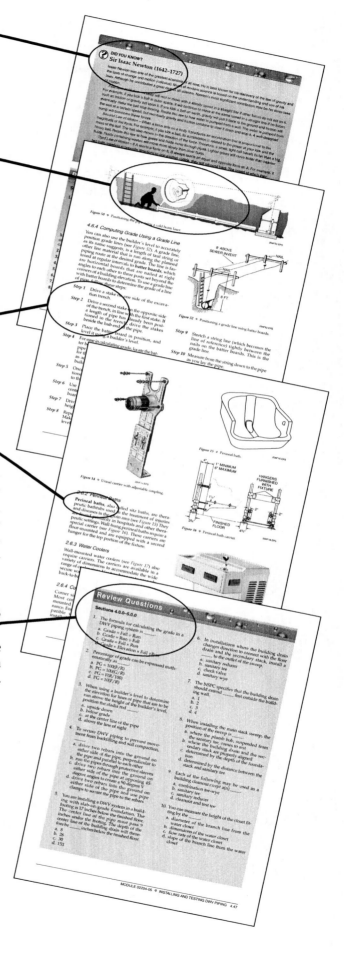

Contents

02301-06 Applied Math 1.i
Introduces trainees to math concepts they will use on the job, including weights and measures, area and volume, temperature, pressure, and force. Also reviews the six simple machines: inclined planes, levers, pulleys, wedges, screws, and wheels and axles. **(17.5 Hours)**

02302-06 Sizing Water Supply Piping 2.i
Teaches techniques for sizing water supply systems, including calculating system requirements and demand, developed lengths, and pressure drops. Also reviews the factors that can reduce efficiency of water supply piping. **(17.5 Hours)**

02303-06 Potable Water Treatment 3.i
Explains how to disinfect, filter, and soften water supply systems. Discusses how to troubleshoot water supply problems, flush out visible contaminants from a plumbing system, and disinfect a potable water plumbing system. **(15 Hours)**

02304-06 Backflow Preventers 4.i
Introduces the different types of backflow prevention devices and discusses how they work, where they are used, and how they are installed. **(20 Hours)**

02305-06 Types of Venting 5.i
Reviews the different types of vents that can be installed in a DWV system and how they work. Also teaches design and installation techniques. **(20 Hours)**

02306-06 Sizing DWV and Storm Systems 6.i
Explains how to calculate drainage fixture units for waste systems. Reviews how to size drain, waste, and vent (DWV) systems; storm drainage systems; and roof storage and drainage systems. **(20 Hours)**

02307-06 Sewage Pumps and Sump Pumps 7.i
Discusses the installation, diagnosis, and repair of pumps, controls, and sumps in sewage and storm water removal systems. **(17.5 Hours)**

02308-06 Corrosive-Resistant Waste Piping 8.i
Discusses corrosive wastes and reviews related safety issues and hazard communications. Discusses how to determine when corrosive-resistant waste piping needs to be installed, as well as how to correctly select and properly connect different types of piping. **(7.5 Hours)**

02309-06 Compressed Air 9.i
Explains the principles of compressed air systems and describes their components and accessories. Reviews installation and periodic servicing of air compressor systems. **(10 Hours)**

Glossary of Trade Terms G.1

Index I.1

Contren® Curricula

NCCER's training programs comprise more than 40 construction, maintenance, and pipeline areas and include skills assessments, safety training, and management education.

Boilermaking
Carpentry
Carpentry, Residential
Cabinetmaking
Concrete Finishing
Construction Craft Laborer
Construction Technology
Core Curriculum: Introductory Craft Skills
Currículum Básico
Electrical
Electrical, Residential
Electrical Topics, Advanced
Electronic Systems Technician
Exploring Careers in Construction
Fundamentals of Mechanical and Electrical Mathematics
Heating, Ventilating, and Air Conditioning
Heavy Equipment Operations
Highway/Heavy Construction
Instrumentation
Insulating
Ironworking
Maintenance, Industrial
Masonry
Millwright
Mobile Crane Operations
Painting
Painting, Industrial
Pipefitting
Pipelayer
Plumbing
Reinforcing Ironwork
Rigging
Scaffolding
Sheet Metal
Site Layout
Sprinkler Fitting
Welding

Pipeline
Control Center Operations, Liquid
Corrosion Control
Electrical and Instrumentation
Field Operations, Liquid
Field Operations, Gas
Maintenance
Mechanical

Safety
Field Safety
Orientación de Seguridad
Safety Orientation
Safety Technology

Management
Introductory Skills for the Crew Leader
Project Management
Project Supervision

Acknowledgments

This curriculum was revised as a result of the farsightedness and leadership of the following sponsors:

Associated Builders and Contractors of Southern California
Ivey Mechanical Company, LLC
JF Ingram
Wat-Kem Mechanical, Inc.

This curriculum would not exist were it not for the dedication and unselfish energy of those volunteers who served on the Authoring Team. A sincere thanks is extended to the following:

Jonathan Byrd
Steve Guy
Charles Owenby
Thomas J. Swafford, CPC
Ray G. Thornton

NCCER PARTNERING ASSOCIATIONS

American Fire Sprinkler Association
American Petroleum Institute
American Society for Training & Development
Associated Builders & Contractors, Inc.
Association for Career and Technical Education
Associated General Contractors of America
Carolinas AGC, Inc.
Construction Industry Institute
Construction Users Roundtable
Design-Build Institute of America
Electronic Systems Industry Consortium
Merit Contractors Association of Canada
Metal Building Manufacturers Association
National Association of Minority Contractors
National Association of State Supervisors for Trade and Industrial Education
National Association of Women in Construction
National Insulation Association
National Ready Mixed Concrete Association
National Systems Contractors Association
National Utility Contractors Association
National Technical Honor Society
North American Crane Bureau
North American Technician Excellence
Painting & Decorating Contractors of America
Portland Cement Association
SkillsUSA
Steel Erectors Association of America
Texas Gulf Coast Chapter ABC
U.S. Army Corps of Engineers
University of Florida
Women Construction Owners & Executives, USA

Plumbing Level Three

02301-06

Applied Math

02301-06
Applied Math

Topics to be presented in this module include:

1.0.0	Introduction	1.2
2.0.0	Weights and Measures	1.2
3.0.0	Measuring Area and Volume	1.5
4.0.0	Temperature, Pressure, and Force	1.14
5.0.0	Simple Machines	1.23
6.0.0	The Worksheet	1.27

Overview

Precise measurements are fundamental to building effective and efficient plumbing systems. Plumbers must be thoroughly familiar with the formulas for measuring area and volume. They should be comfortable converting decimals to fractions and back, as well as converting between the English system of measurement (the common U.S. system) and the metric system used worldwide. Area often needs to be determined for floors and roofs, and volume calculation is a part of nearly all plumbing jobs. When installing pipes and fixtures, plumbers also need to compute the weight of water for a given volume (the gallon capacity) to ensure that pipes and fittings are adequately supported.

Plumbers use many of the physical laws of temperature in their work. These include the rule that energy seeks equilibrium, the concept of conduction (the transfer of heat from warmer to cooler objects), and the principle of thermal expansion (materials expand when they are heated and contract when they are cooled). Thermal expansion occurs in all dimensions of an object and alters the area and volume of the heated material. Heat flow and thermal expansion are both major concerns of plumbers when installing a plumbing system.

Pressure is related to temperature, because increasing temperature increases pressure. Compression of a liquid or gas also exerts increased pressure on the surrounding environment. Being able to determine water and air pressure is another important skill. All these are essential concepts in plumbing, and plumbers work with them every day.

Focus Statement
The goal of the plumber is to protect the health, safety, and comfort of the nation job by job.

Code Note
Codes vary among jurisdictions. Because of the variations in code, consult the applicable code whenever regulations are in question. Referring to an incorrect set of codes can cause as much trouble as failing to reference codes altogether. Obtain, review, and familiarize yourself with your local adopted code.

Portions of this publication reproduce tables and figures from the *2003 International Plumbing Code* and the *2000 IPC Commentary*, International Code Council, Inc., Falls Church, Virginia. Reproduced with permission. All rights reserved.

Objectives

When you have completed this module, you will be able to do the following:

1. Identify the weights and measures used in the English and metric systems.
2. Describe how to calculate area and volume.
3. Describe the practical applications of area and volume in plumbing.
4. Explain the concepts of temperature and pressure and how they apply to plumbing installations.
5. Explain the functions and applications of six simple machines: inclined plane, lever, pulley, wedge, screw, and wheel and axle.

Trade Terms

Applied mathematics
Area
Bimetallic thermometer
Celsius scale
Centigrade scale
Circle
Conduction
Cube
Cubic foot
Cubic meter
Cylinder
Decimal of a foot
Electrical thermometer
English system
Equilibrium
Fahrenheit scale
Fulcrum
Gallon
Head
Inclined plane
Isosceles triangle
Kelvin scale
Lever
Liquid thermometer
Liter
Metric system
Pounds per square inch
Pressure
Prism
Pulley
Rectangle
Right triangle
SI system
Screw
Simple machine
Square
Square foot
Square meter
Temperature
Thermal expansion
Thermometer
Volume
Water Supply Fixture Unit (WSFU)
Wedge
Wheel and axle
Work

Required Trainee Materials

1. Appropriate personal protective equipment
2. Sharpened pencils and paper
3. Copy of local applicable code
4. Calculator

Prerequisites

Before you begin this module, it is recommended that you successfully complete *Core Curriculum; Plumbing Level One; Plumbing Level Two.*

This course map shows all of the modules in the third level of the *Plumbing* curriculum. The suggested training order begins at the bottom and proceeds up. Skill levels increase as you advance on the course map. The local Training Program Sponsor may adjust the training order.

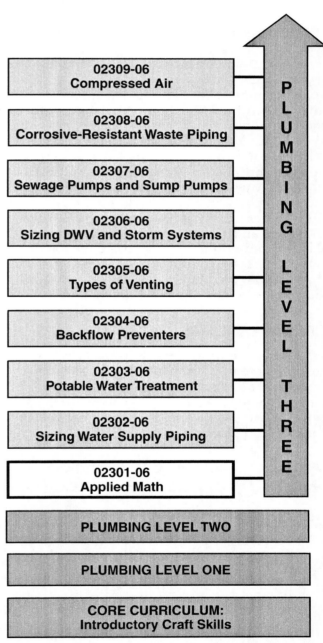

1.0.0 ◆ INTRODUCTION

Plumbers use math every day. However, that doesn't mean you have to be a professional mathematician to install a perfectly functioning plumbing system. You may not realize you are using math when you size a drain, level a water heater, or calculate a grade for a run of pipe, but you are. This module will show you the math behind common plumbing activities.

Plumbers use a special branch of math called **applied mathematics**. Applied mathematics is any mathematical process that you use to accomplish a specific task. It helps you get a job done correctly. It enables you to find the length of a run of pipe, the slope of a drain, the **area** of a floor, the volume of a climb-in interceptor, the **temperature** of wastewater—anything where you need to find a precise number. Plumbers use formulas and calculations to find those numbers. You need to understand how formulas and calculations work so that you can use them, too.

In this module, you will learn about the weights and measures that plumbers use. You will learn ways to calculate the size of flat surfaces and the amount of space inside a container. You will also learn about heat and other forces that act on liquids and gases in containers, and you will discover how plumbers use basic machines to move things. By learning these things, you will improve your plumbing skills, and the results will show in the quality of work you do.

2.0.0 ◆ WEIGHTS AND MEASURES

Systems of measurement allow people to exchange information about all kinds of numbers. Whether it's the angle of a pipe, the amount of water in a tank, the capacity of a pump, or the distance to the sewer line, plumbers are always thinking and working in terms of how much or how little, how large or how small, and how long or how short. Systems of measurement are essential to the plumbing trade.

Plumbers use two systems of measurement when working with plumbing installations. The next section will review the two major systems of weights and measures and how to convert measurements from one system to the other.

2.1.0 The English System

The system of weights and measures currently used in the United States is called the **English system**. It is a version of a system created in England many centuries ago called the British Imperial system. The system was brought over by the European settlers who first established colonies in North America. The system has gradually developed over time to its current form. Some measures and values differ from the original British Imperial system. Currently, the United States is the only country that still uses the English system for all its weights and measures.

Most of the pipe, tools, fixtures, fittings, and other plumbing and construction equipment in the United States are sized according to the English system. Many people in the United States are familiar with the weights and measures included in the English system. However, the manner in which the different measures work together (such as how many feet are in a yard) is not always known or understood. *Table 1* lists weights and measures that are commonly used by plumbers.

2.2.0 The Metric (SI) System

The **metric system** was developed in France in the eighteenth century. It has since become a worldwide standard in almost all professional and scientific fields. The metric system is also called the

DID YOU KNOW?
The Standardization of Measurement Systems

The science of the standardization of measures is called metrology. Beginning in the late 1880s, many countries established laboratories to develop and maintain standards for weights and measures. Two of the most famous of these standards laboratories are the National Institute of Standards and Technology (NIST) in the United States and the National Physical Laboratory (NPL) in the United Kingdom. During the twentieth century, these laboratories developed guidelines for an international system of weights and measures. Since then, most countries have adopted these guidelines. The International Bureau of Weights and Measures, located in France, is now responsible for establishing and maintaining international standards. The other standards laboratories help implement these standards in their own countries.

Table 1 English System—Common Weights and Measures

English System—Common Weights and Measures		
Linear Measures		
1 foot	=	12 inches
1 yard	=	36 inches
1 yard	=	3 feet
1 mile	=	5,280 feet
1 mile	=	1,760 yards
Area Measures		
1 square foot	=	144 square inches
1 square yard	=	9 square feet
1 acre	=	43,560 square feet
1 square mile	=	640 acres
Volume Measures		
1 cubic foot	=	1,728 cubic inches
1 cubic yard	=	27 cubic feet
1 freight ton	=	40 cubic feet
1 register ton	=	100 cubic feet
Liquid Volume Measures		
1 gallon	=	128 fluid ounces
1 gallon	=	4 quarts
1 barrel	=	31.5 gallons
1 petroleum barrel	=	42 gallons
Weights		
1 pound	=	16 ounces
1 hundredweight	=	100 pounds
1 ton	=	2,000 pounds
Pressure Measures		
1 pound per square inch absolute	=	14.7 pounds at sea level
1 pound per square inch gauge	=	Varies with altitude

SI system. SI stands for *Système International d'Unités,* or International System of Units.

One of the major advantages of the metric system is that it is rigorously standardized. This makes it easy to memorize the basic system. There are only seven basic units of measurement in the modern metric system (see *Table 2*). All other metric measures result from various combinations of these basic units. Multiples and fractions of the basic units are expressed as powers of 10 (except for seconds, of which there are 60 in a minute). A standard set of prefixes is used to denote these larger or smaller numbers (refer to *Table 2*). That way, you will always know that a kilometer is 1,000 meters just by looking at the prefix. Note that the basic measure of mass, the kilogram, is already a multiple. A kilogram is 1,000 grams. Because a gram is such a small amount of mass, however, a larger base unit seemed more appropriate.

DID YOU KNOW?
History of the Metric System

The idea of basing weights and measures on a decimal system (numbers that are multiples of 10) was first proposed in the sixteenth century by the Dutch engineer Simon Stevin. In 1790, Thomas Jefferson proposed that the new United States develop a decimal weights and measures system. The same year, King Louis XVI of France instructed scientists in his country to develop such a system. Five years later, France adopted the metric system. In 1875, France hosted the Convention of the Metre. Eighteen nations, including the United States, signed a treaty that established an international body to govern the development of weights and measures. In 1960, this organization officially changed the name of the metric system to the International System of Units (SI).

The metric system was designed with three goals in mind:

a. The system should use units based on unchanging quantities in nature.
b. All measurement units should derive from only a few base units.
c. The system should use multiples of 10.

The modern metric system still follows these three simple guidelines.

Originally, the meter was measured as one ten-millionth the distance from the North Pole to the equator. Bars of that length were constructed in brass and later in platinum and sent to the standards laboratories of participating countries. Scientists then learned that the actual distance from the pole to the equator was different from the measurement the French scientists used. The length of the meter was recalculated using more precise methods, including the distance traveled by a light beam in a fraction of a second.

In 1994, the United States government required that consumer products feature measurements in both SI and English form. Government-issued construction contracts now use metric measurements. With the advent of the *International Plumbing Code®* and other international standards, the United States may eventually follow the rest of the world and completely adopt the SI system of weights and measures.

Table 2 Basic Measures of the Metric System

The unit of...	Is called the...
Length	Meter
Mass	Kilogram
Temperature	Kelvin
Time	Second
Pressure	Pascal
Electric current	Ampere
Light intensity	Candela
Substance amount	Mole

The prefix...	Means...
Micro-	One millionth
Milli-	One thousandth
Centi-	One hundredth
Deci-	One tenth
Deka-	Ten times
Hecto-	One hundred times
Kilo-	One thousand times
Mega-	One million times

CAUTION

Always make sure that you express measurements using the correct system. Errors caused by using the wrong system of measurement can cost time and money. Never use a metric tool on English system pipe and fittings, or vice versa. You could damage both the tool and the fitting.

With the growing popularity of international standards, such as the *International Plumbing Code®*, plumbers will frequently encounter metric weights and measures. The federal government uses metric measurements in its contracts, so plumbers working on a federal government project will need to understand metric weights and measures. And as more and more companies conduct their business internationally, they adopt metric measurements as well.

2.3.0 Converting Measurements

When working with systems of measurement, you may be required to convert decimals to fractions. Converting decimals of a foot to fractions is a two-step process. First, multiply the decimal by 12, which is the number of inches in a foot. Then, multiply that number by the base of the fraction you are seeking. For example, multiply by 16 if you are looking for sixteenths of an inch, by 8 if you are looking for eighths of an inch, and so on.

Another important type of conversion is between the English and metric systems. You've probably seen tools marked with lengths in both inches and millimeters or fixtures with shipping weights in both pounds and kilograms. However, there may be times in the field when you have to perform a conversion calculation yourself. You should become familiar with the basic conversions between the English and metric systems.

Appendix A provides a reference for converting the most common weights and measures. To convert a measurement into its equivalent in the other system, multiply the measurement by the number in the far right column. For example, you want to know how many kilometers equal 5 miles. Referring to the table, you see that the number must be multiplied by 1.6. The result, therefore, is that 5 miles equal 8 kilometers.

Sometimes a conversion requires two calculations, one within a system and one between systems. For example, to convert from feet to meters, where you do not have an exact conversion factor, you would first use the factor for converting feet to centimeters (30.48) and then divide the result by 100 to convert from centimeters to meters. Practice calculating some conversions of your own, referring to *Table 2* if needed to convert within the metric system.

2.3.1 Study Problems: English-Metric Conversion

Refer to the conversion formulas in *Appendix A*. Round off answers to three decimal places.

1. Convert 432 square inches to square centimeters.
   Answer: _____

2. Convert 124 yards to meters.
   Answer: _____

3. Convert 1,240 grams to ounces.
   Answer: _____

4. Convert 450 psi to kilopascals.
   Answer: _____

5. Convert 0.5 pints to liters.
   Answer: _____

6. Convert 75 kilopascals to psi.
   Answer: _____

Review Questions

Sections 1.0.0–2.0.0

1. A mile consists of 5,280 feet or _____ yards.
   a. 52,800
   b. 1,760
   c. 43,560
   d. 1,728

2. One register ton equals _____ cubic feet.
   a. 27
   b. 40
   c. 100
   d. 160

3. The prefix *centi-* means _____.
   a. one hundredth
   b. one millionth
   c. one hundred times
   d. one thousand times

4. The prefix meaning *one tenth* is _____.
   a. deci-
   b. deka-
   c. hecto-
   d. micro-

5. The basic unit of light intensity in the metric system is the _____.
   a. kelvin
   b. ampere
   c. mole
   d. candela

---

7. Convert 1,331 cubic inches into cubic meters.
   Answer: _____

8. Convert 3,700 pounds to megagrams (metric tons).
   Answer: _____

9. Convert 4,350 cubic centimeters to cubic feet.
   Answer: _____

10. Convert 1,500 milliliters to gallons.
    Answer: _____

3.0.0 ♦ MEASURING AREA AND VOLUME

Plumbers must know how to measure the size of flat surfaces and the space inside different types of containers. Typical flat surfaces include building foundations, floors, roofs, and parking lots. Pipes are probably the most common container you will work with as a plumber. Swimming pools, interceptors, and lavatories are also types of containers.

In this section, you will learn how to measure flat surfaces and the space inside containers. With practice, these measurements will become as familiar to you as any other tool you use on the job, like a wrench or a plumb bob. *Appendix B* offers a quick reference to the basic formulas for area and volume.

3.1.0 Measuring Area

Area is the measurement of a flat surface (see *Figure 1*). Area is measured using two dimensions: length and width. Floor space is a common type of area that plumbers encounter on the job. You have already learned that plumbers must take floor space into account when designing plumbing installations. In this section you will learn how to calculate the area of floor spaces and other areas.

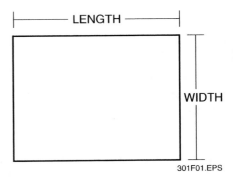

Figure 1 ♦ The concept of area.

In the English system, area is expressed in **square feet**. You will also see this written as sq. ft. or ft². In the metric system, area is measured in **square meters**. This can also be written as m².

Length and width are measured in **decimals of a foot**. You learned earlier in this curriculum that a decimal of a foot is a decimal fraction in which the denominator is 12, instead of the usual 10 (see *Table 3*).

You can also convert inches to their decimal equivalent in feet without the table of decimal equivalents. All you need to do is some simple division. First, express the inches as a fraction that has 12 as the denominator. You use 12 because there are 12 inches in a foot. Then reduce the fraction and convert it to a decimal.

Consider the following example. To find the decimal equivalent in feet of 3 inches, follow these three steps:

Step 1 Express 3 inches as a fraction with 12 as the denominator.

$\frac{3}{12}$

Step 2 Reduce the fraction.

$\frac{3}{12} = \frac{1}{4}$

Step 3 Convert the reduced fraction to a decimal. You can do this by dividing the denominator, 4, into 1.00:

$1.00 \div 4 = 0.25$

Thus, 3 inches converts to 0.25 of a foot. For more complicated fractions, use a calculator.

In plumbing, it will be useful for you to know how to calculate the area of three basic shapes:

- Rectangles
- Right triangles
- Circles

With this knowledge, you will be able to calculate most of the types of areas you will encounter on the job. The following sections review how to calculate the area of rectangles, right triangles, and circles.

3.1.1 Measuring the Area of a Rectangle

A **rectangle** (*Figure 2A*) is a four-sided figure in which all four corners are right angles. To calculate the area (A) of a rectangle, multiply the length (l) by the width (w). Expressed as a formula, the calculation is:

$A = lw$

A **square** is a special type of rectangle in which all four sides are the same length. Use the same formula to determine the area of a square (see *Figure 2B*). You can also calculate the area of a square by squaring the length of one of the sides:

$A = s^2$

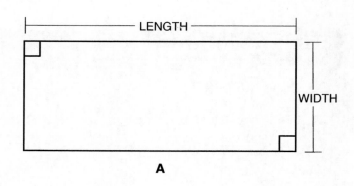

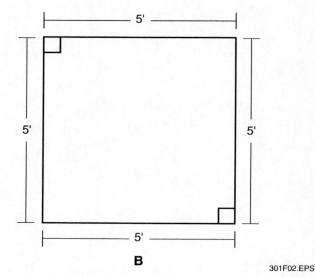

Figure 2 ♦ Measuring the area of a rectangle and a square.

Let's try this formula on a rectangle that is 5 feet 1½ inches by 8 feet 10¾ inches:

Step 1 First, convert the inches to decimals of a foot.

5 feet 1½ inches = 5.1250 feet
8 feet 10¾ inches = 8.895833 feet

Step 2 Now round your answer to three decimal places, ensuring that round-ups are accurate. When rounding numbers, any number followed by 5 through 9 is rounded up, while any number followed by 0 through 4 remains the same.

5.1250 feet = 5.125 feet
8.895833 feet = 8.896 feet

Note that the number of decimal places you round to depends on the job being done and the desired level of accuracy. When in doubt, ask your supervisor.

Table 3 Inches Converted to Decimals of a Foot

Inches	Decimals of a Foot	Inches	Decimals of a Foot	Inches	Decimals of a Foot	Inches	Decimals of a Foot
1/16	0.005	3 1/16	0.255	6 1/16	0.505	9 1/16	0.755
1/8	0.010	3 1/8	0.260	6 1/8	0.510	9 1/8	0.760
3/16	0.016	3 3/16	0.266	6 3/16	0.516	9 3/16	0.766
1/4	0.021	3 1/4	0.271	6 1/4	0.521	9 1/4	0.771
5/16	0.026	3 5/16	0.276	6 5/16	0.526	9 5/16	0.776
3/8	0.031	3 3/8	0.281	6 3/8	0.531	9 3/8	0.781
7/16	0.036	3 7/16	0.286	6 7/16	0.536	9 7/16	0.786
1/2	0.042	3 1/2	0.292	6 1/2	0.542	9 1/2	0.792
9/16	0.047	3 9/16	0.297	6 9/16	0.547	9 9/16	0.797
5/8	0.052	3 5/8	0.302	6 5/8	0.552	9 5/8	0.802
11/16	0.057	3 11/16	0.307	6 11/16	0.557	9 11/16	0.807
3/4	0.063	3 3/4	0.313	6 3/4	0.563	9 3/4	0.813
13/16	0.068	3 13/16	0.318	6 13/16	0.568	9 13/16	0.818
7/8	0.073	3 7/8	0.323	6 7/8	0.573	9 7/8	0.823
15/16	0.078	3 15/16	0.328	6 15/16	0.578	9 15/16	0.828
1	0.083	4	0.333	7	0.583	10	0.833
1 1/16	0.089	4 1/16	0.339	7 1/16	0.589	10 1/16	0.839
1 1/8	0.094	4 1/8	0.344	7 1/8	0.594	10 1/8	0.844
1 3/16	0.099	4 3/16	0.349	7 3/16	0.599	10 3/16	0.849
1 1/4	0.104	4 1/4	0.354	7 1/4	0.604	10 1/4	0.854
1 5/16	0.109	4 5/16	0.359	7 5/16	0.609	10 5/16	0.859
1 3/8	0.115	4 3/8	0.365	7 3/8	0.615	10 3/8	0.865
1 7/16	0.120	4 7/16	0.370	7 7/16	0.620	10 7/16	0.870
1 1/2	0.125	4 1/2	0.375	7 1/2	0.625	10 1/2	0.875
1 9/16	0.130	4 9/16	0.380	7 9/16	0.630	10 9/16	0.880
1 5/8	0.135	4 5/8	0.385	7 5/8	0.635	10 5/8	0.885
1 11/16	0.141	4 11/16	0.391	7 11/16	0.641	10 11/16	0.891
1 3/4	0.146	4 3/4	0.396	7 3/4	0.646	10 3/4	0.896
1 13/16	0.151	4 13/16	0.401	7 13/16	0.651	10 13/16	0.901
1 7/8	0.156	4 7/8	0.406	7 7/8	0.656	10 7/8	0.906
1 15/16	0.161	4 15/16	0.411	7 15/16	0.661	10 15/16	0.911
2	0.167	5	0.417	8	0.667	11	0.917
2 1/16	0.172	5 1/16	0.422	8 1/16	0.672	11 1/16	0.922
2 1/8	0.177	5 1/8	0.427	8 1/8	0.677	11 1/8	0.927
2 3/16	0.182	5 3/16	0.432	8 3/16	0.682	11 3/16	0.932
2 1/4	0.188	5 1/4	0.438	8 1/4	0.688	11 1/4	0.938
2 5/16	0.193	5 5/16	0.443	8 5/16	0.693	11 5/16	0.943
2 3/8	0.198	5 3/8	0.448	8 3/8	0.698	11 3/8	0.948
2 7/16	0.203	5 7/16	0.453	8 7/16	0.703	11 7/16	0.953
2 1/2	0.208	5 1/2	0.458	8 1/2	0.708	11 1/2	0.958
2 9/16	0.214	5 9/16	0.464	8 9/16	0.714	11 9/16	0.964
2 5/8	0.219	5 5/8	0.469	8 5/8	0.719	11 5/8	0.969
2 11/16	0.224	5 11/16	0.474	8 11/16	0.724	11 11/16	0.974
2 3/4	0.229	5 3/4	0.479	8 3/4	0.729	11 3/4	0.979
2 13/16	0.234	5 13/16	0.484	8 13/16	0.734	11 13/16	0.984
2 7/8	0.240	5 7/8	0.490	8 7/8	0.740	11 7/8	0.990
2 15/16	0.245	5 15/16	0.495	8 15/16	0.745	11 15/16	0.995
3	0.250	6	0.500	9	0.750	12	1.000

Step 3 Multiply the length by the width.
A = lw
A = (5.125)(8.896)
A = 45.592 sq. ft.

The total area of a rectangle that is 5 feet 1½ inches by 8 feet 10¾ inches, therefore, is 45.592 square feet.

3.1.2 Measuring the Area of a Right Triangle

A **right triangle** is a three-sided figure that has an internal angle that equals 90 degrees. The area of a right triangle is one-half the product of the base (b) and height (h). In other words:

A = ½(bh)

An **isosceles triangle** is a special kind of triangle that has two sides of equal length. Use the same formula to calculate the area of an isosceles triangle (see *Figure 3*).

Test this formula on a triangle where the height is 5 feet 2¾ inches and the base is 6 feet 6½ inches. Calculate the area for this triangle by following these steps:

Step 1 Convert the inches into decimals of a foot.
5 feet 2¾ inches = 5.229166 feet
6 feet 6½ inches = 6.541666 feet

Step 2 Now round your answer to three decimal places, ensuring that round-ups are accurate.
5.229166 feet = 5.229 feet
6.541666 feet = 6.542 feet

Step 3 Determine the area of the triangle.
A = ½(bh)
A = (½)(6.542)(5.229)
A = (½)(34.208)
A = 17.104 sq ft

The area of a right triangle whose height is 5 feet 2¾ inches and base is 6 feet 6½ inches long, therefore, is 17.104 square feet.

Sometimes, you might not know the length of one of the legs in a right triangle. To calculate the missing leg, you can use the Pythagorean theorem. The Pythagorean theorem states the following:

$$a^2 + b^2 = c^2$$

Refer to *Figure 3*. You see that a and b are the two sides that meet at the 90-degree angle. The side marked c is the hypotenuse of the triangle. To find either a or b, use the following calculations:

$$a = \sqrt{c^2 - b^2}$$
$$b = \sqrt{c^2 - a^2}$$

Use the square root function on your calculator to obtain the square root of a, b, and c. Most calculators have a square root key (see *Figure 4*). It is marked with the √ symbol. To use it, first enter the number for which you want to calculate the square root. Then press the square root key. The calculator will display the square root of the number you entered.

If the number you entered represented feet, then the decimals must be converted to inches and fractions of an inch. If the number you entered was in inches, the decimals must be converted to fractions of an inch.

3.1.3 Measuring the Area of a Circle

A **circle** is a one-sided figure where any point on the side is the same distance from the center. The strainer on a can wash drain is an example of a circular surface you will encounter as a plumber. The area of a circle is a little more complicated to calculate than that of a rectangle or triangle. To calculate the area of a circle, you will use the radius of the circle and pi. The radius of a circle is one-half the diameter (which is formed by a straight

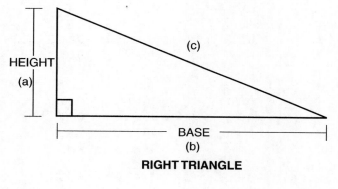
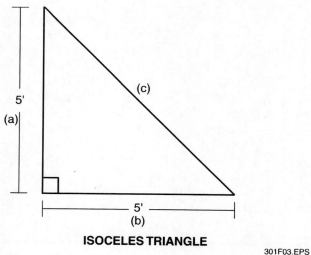

Figure 3 ◆ Right triangles.

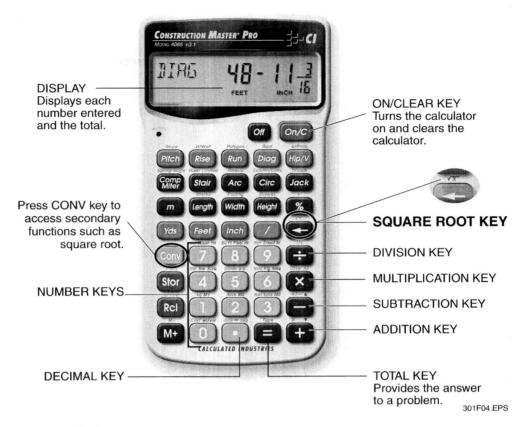

Figure 4 ◆ Square root key on a calculator.

line through the center of the circle to its edges). Pi is equal to 3.1416 and is usually represented by the Greek letter π. First, multiply the radius (r) by itself (this is called squaring). Then, multiply the result by pi.

$A = \pi r^2$

To find the area of a circle with a diameter of 4 feet 3½ inches, for example, follow these steps:

Step 1 The radius is one-half the diameter, or 2 feet 1¾ inches. Now convert the inches to decimals of a foot.

2 feet 1¾ inches = 2.145833 feet

Step 2 Now round your answer to three decimal places, ensuring that round-ups are accurate.

2.145833 feet = 2.146 feet

Step 3 Use the formula to calculate the area. Do not round pi.

$A = \pi r^2$
$A = (3.1416)(2.146^2)$
$A = (3.1416)(4.605)$
$A = 14.47$ sq. ft.

You have shown that a circle with a diameter of 4 feet 3½ inches has an area of 14.47 square feet.

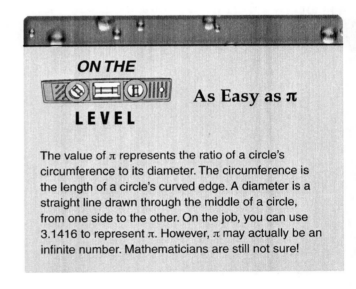

ON THE LEVEL — As Easy as π

The value of π represents the ratio of a circle's circumference to its diameter. The circumference is the length of a circle's curved edge. A diameter is a straight line drawn through the middle of a circle, from one side to the other. On the job, you can use 3.1416 to represent π. However, π may actually be an infinite number. Mathematicians are still not sure!

3.1.4 Area Calculations in Plumbing

You have learned how to calculate the area of different shapes. With this knowledge, you can calculate the area of almost any floor plan. Always refer to the scale in a construction drawing to determine the proper measurements of the floor.

On the job, you may have to calculate the area of a space shaped like a rectangle or a right triangle. Let's say you have to install an area drain in a parking lot that is 30 feet 6½ inches by 50 feet 3 inches. The parking lot's size affects the amount of storm water runoff that the drain will have to handle. This factor determines the size of the area drain you select.

Step 1 First, convert the inches to decimals of a foot.

30 feet 6½ inches = 30.541666 feet
50 feet 3 inches = 50.250 feet

Step 2 Now round each answer to three decimal places, ensuring that round-ups are accurate.

30.541666 feet = 30.542 feet
50.250 feet = 50.250 feet

Step 3 Multiply the length by the width.

A = lw
A = (30.542)(50.250)
A = 1,534.74 sq ft

The total area of a parking lot that is 30 feet 6½ inches by 50 feet 3 inches is 1,534.74 square feet.

When calculating the area of a surface that is not a simple rectangle or right triangle, break the floor plan into smaller areas shaped like rectangles and right triangles (see *Figure 5*). Then calculate the area for each of the smaller areas using the appropriate formula. Finally, add all the areas together. You will then know the total overall area of the floor.

3.2.0 Measuring Volume

Volume is a measure of capacity, the amount of space within something (see *Figure 6*). Volume is measured using three dimensions: length, width, and height. Plumbers must be able to calculate the volumes of various spaces accurately to ensure the efficient operation of plumbing installations. Some of the volumes that plumbers need to be able to calculate include:

- Drain receptors
- Fresh water pipes
- Drain, waste, and vent (DWV) pipes
- Grease and oil interceptors
- Water heaters
- Septic tanks
- Swimming pools
- Catch basins

In the English system, space is measured in **cubic feet** (abbreviated as cu. ft. or ft^3). The measure of liquid volume is **gallons** (abbreviated as gal.). There are 7.48 gallons in 1 cubic foot. In the metric system, the measure of space is **cubic meters** (abbreviated as m^3) and the measure of liquid is in **liters** (abbreviated as L). There are 1,000 liters in a cubic meter.

You can construct three-dimensional spaces by combining flat surfaces like rectangles, squares, right triangles, and circles. When two parallel rectangles, squares, or right triangles are connected by rectangles, the space is called a **prism**. A tube with a circular cross section is called a **cylinder**.

3.2.1 Measuring the Volume of a Rectangular Prism

Rectangular prisms are three-dimensional spaces where rectangles make up the sides, top, and bottom (see *Figure 7*). A 2 × 4 piece of lumber is a rectangular

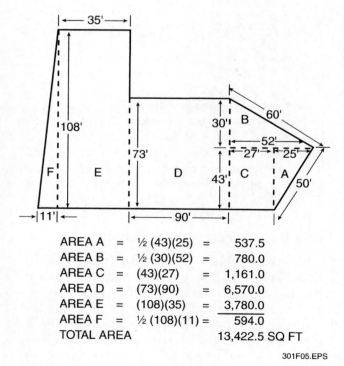

Figure 5 ◆ Subdividing an area into simple shapes.

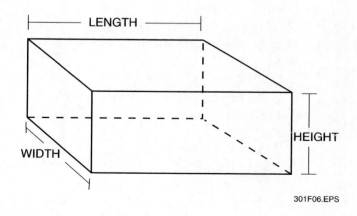

Figure 6 ◆ The concept of volume.

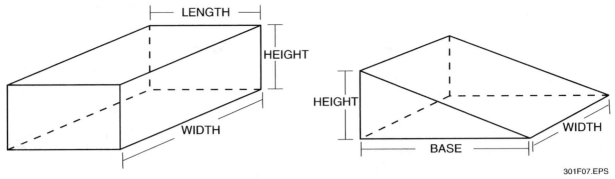

Figure 7 ◆ Rectangular prism and right triangular prism.

prism, as is the interior of a truck bed. To calculate the volume of a rectangular prism, multiply the length by the width by the height.

For example, assume you have a rectangle with a length of 5 feet 4¾ inches, a width of 2 feet 1½ inches, and a height of 6 feet. Calculate the volume this way:

Step 1 Convert the inches into decimals of a foot.

5 feet 4¾ inches = 5.395833 feet
2 feet 1½ inches = 2.1250 feet

Step 2 Now round your answer to three decimal places, ensuring that round-ups are accurate.

5.395833 feet = 5.396 feet
2.1250 feet = 2.125 feet

Step 3 Calculate the volume using the formula:

V = lwh
V = (5.396)(2.125)(6)
V = (11.467)(6)
V = 68.802 cu ft

The volume of the given rectangle is 68.802 cubic feet.

A **cube** is a rectangular prism where the length, width, and height are all the same. To find the volume of a cube, use the formula for finding the volume of a rectangular prism.

3.2.2 Measuring the Volume of a Right Triangular Prism

The volume of a right triangular prism is ½ times the product of the length, width, and height (refer to *Figure 7*). For example, if you had a right triangle prism with a base and width of 3 feet and a height of 5 feet, the volume would be calculated as follows:

V = ½(bwh)
V = ½(3)(3)(5)
V = ½(9)(5)
V = ½(45)
V = 22.5 cu ft

The volume of the right triangular prism is 22.5 cubic feet.

3.2.3 Measuring the Volume of a Cylinder

A cylinder is a straight tube with a circular cross section (see *Figure 8*). It is a common plumbing shape. Pipes are long, narrow cylinders. Many water heaters are also cylindrical. You can calculate the volume in cubic feet of a cylinder by multiplying the area of the circular base (πr^2) by the height (h).

In the following example, a cylinder has an inside diameter of 30 feet and a height of 50 feet 9 inches. Obtain the volume in cubic feet using the following steps:

Step 1 The ends of a cylinder are circles. So first, you must determine the area of the circle. In the example, the radius (half the diameter) is 15 feet.

A = πr^2
A = (3.1416)(15²)
A = (3.1416)(225)
A = 706.86 sq ft

Step 2 To calculate the volume of the tank in cubic feet, multiply the area of the circle by the height. Remember to convert the measurement into decimals of a foot.

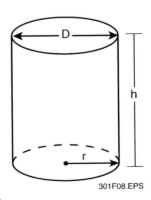

Figure 8 ◆ Cylinder.

$$V = Ah$$
$$V = (706.86)(50.75)$$
$$V = 35{,}873.145 \text{ cu ft}$$

The volume of a tank with an inside diameter of 30 feet and a height of 50 feet 9 inches is 35,873.145 cubic feet.

3.2.4 Volume Calculations in Plumbing

For manufactured items, you can usually find the dimensions and volume in the manufacturer's specifications. For site-built containers, such as large grease interceptors, you will have to calculate the volume yourself. Don't forget to calculate the volume of objects inside a container, such as baffles, fittings, or even fixtures. Calculate their volumes and subtract them from the overall volume of the space (see *Figure 9*).

When determining volume, it is also important to consider weight. Take, for example, a swimming pool installed on a patio. The larger the pool, the more water it will hold and the heavier the water inside it will be. Consequently, the deck will require more support to hold the pool. When installing pipes and fixtures, plumbers must be able to calculate the weight of water for a given volume to ensure that the pipe or fixture is adequately supported. Local codes provide requirements on proper support.

Once you have calculated the volume of a container such as a pipe, water tank, slop sink, or interceptor, you must perform one more step to determine its gallon capacity. One gallon of water weighs 8.33 pounds. There are 7.48 gallons in a cubic foot. To find the capacity in gallons of a volume measured in cubic feet, multiply the cubic footage by 7.48. If you multiply the weight of 1 gallon of water by the number of gallons in a cubic foot, you will find that a cubic foot of water weighs 62.4 pounds. Remember that these weights are for pure water. Wastewater will have other liquids and solids in it. This means the weight of a volume of wastewater may be greater than that of fresh water. To convert measurements in cubic inches to gallons, divide the cubic inch measurement by 231.

When installing water supply lines, using a smaller pipe size can save the customer both energy and money. You can calculate the difference using the above formula. For example, a ¾-inch-diameter pipe has a cross-sectional area of 0.442 square inches, and a ½-inch-diameter pipe has a cross-sectional area of 0.196 square inches (*Figure 10*). If you perform the calculations, you will find that a ¾-inch pipe will hold 2¼ times the amount of water that a ½-inch pipe will hold. With a larger pipe, a faucet will have to run longer to get hot water to a sink. In addition, the amount of hot water that will be left in the pipe to cool once the faucet is turned off is much greater with a larger pipe. Always consult local codes to determine the water- and energy-saving measures possible for each job.

You can calculate the volume of a complex shape by calculating the volume of its various parts. For example, divide the shape in *Figure 11* into a half cylinder and a rectangular solid.

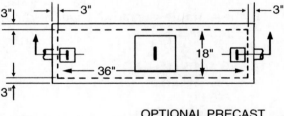

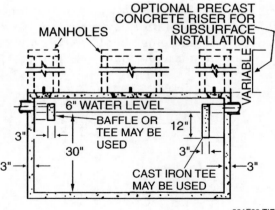

Figure 9 ♦ Calculating the volume of a site-built grease interceptor.

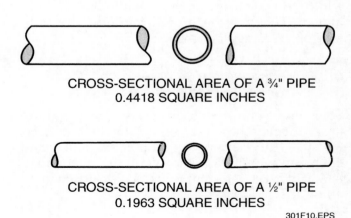

Figure 10 ♦ Cross-sectional areas of pipe.

ON THE LEVEL

Alternate Method to Calculate Cylinder Volume and Gallon Capacity

A shortcut to determine the volume of a cylinder uses the diameter rather than the radius. First, square the diameter of the circular base (d^2). Then, multiply the result by 0.7846 and, finally, multiply by the length (l). The formula is:

$V = (d^2)(0.7854)l$

To find the gallon capacity of a cylinder, use the result of the above calculation in either of the following formulas:

When cylinder measurement is in cubic feet: $G = V(7.48)$
When cylinder measurement is in cubic inches: $G = V \div 231$

The volume of the rectangular portion is found as follows:

$V = lwh$
$V = (12)(6)(8)$ feet
$V = 576$ cu ft

The volume of the half cylinder is found as follows:

$V = (\pi r^2 h) \div 2$
$V = ([3.1416][3][3][8]) \div 2$
$V = (226.1952) \div 2$
$V = 113.098$ cu ft

Now add the two together to get the total volume:

$V = 576$ cu ft $+ 113.098$ cu ft
$V = 689.098$ cu ft

The total volume of the complex shape is 689.098 cubic feet.

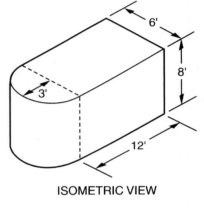

ISOMETRIC VIEW

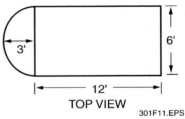

TOP VIEW

Figure 11 ♦ Complex volume.

Review Questions

Section 3.0.0

Remember, when necessary, to round your answers to three decimal places.

1. Seven inches expressed in decimals of a foot is _____.
   a. 84
   b. 0.583
   c. 0.84
   d. 0.292

2. The area of a rectangle with a length of 3 feet and a width of 6 feet is _____.
   a. 9 square feet
   b. 36 square feet
   c. 18 square feet
   d. 8.48 square feet

3. The area of a circle with a diameter of 4 inches is _____.
   a. 16 square inches
   b. 157.974 square inches
   c. 33.333 square inches
   d. 12.566 square inches

4. The volume of a right triangular prism with a base of 2 feet, a width of 5 feet, and a height of 7 feet is _____.
   a. 70 cubic feet
   b. 14 cubic feet
   c. 35 cubic feet
   d. 122 cubic feet

5. The volume of a cylinder with a diameter of 30 inches and a height of 60 inches is _____.
   a. 42,411.60 cubic inches
   b. 169,646.40 cubic inches
   c. 13,324.037 cubic inches
   d. 706.86 cubic inches

4.0.0 ◆ TEMPERATURE, PRESSURE, AND FORCE

The previous section showed how to calculate the space inside three-dimensional containers like pipes, interceptors, and swimming pools. Those calculations are an important step when installing a plumbing system. But once a plumbing system is in place, it has to contain and move liquids and gases. Liquids and gases have properties that affect how they behave inside the system. In this section, you will learn about three of the most important of these properties: temperature, **pressure**, and force.

Temperature, pressure, and force are very important concepts in plumbing. Pipes, fittings, and fixtures are all designed to work within a range of acceptable temperatures and pressures. If those limits are exceeded, the system could be severely damaged, and personal injury could result. A plumbing system that contains materials not appropriate for its operating conditions is also in danger of failing and causing damage and injury. Every plumbing system is designed from the outset with these considerations in mind, so every plumber needs to know the principles that govern temperature and pressure.

4.1.0 Temperature

Temperature is a measure of the heat of an object according to a scale. Heat transfers from warmer objects to cooler ones. This is because heat is a form of energy, and energy seeks **equilibrium**. Equilibrium is a condition in which the temperature is the same throughout the object or space. This is why you get hot when you are working next to a steam pipe; your body is absorbing some of the heat energy of the pipe. The flow of energy from a hotter object to a cooler one is called **conduction**.

Plumbers are very concerned with heat flow when they install a plumbing system. Pipes can lose heat into the surrounding atmosphere, into a nearby wall or ceiling, or into the ground. When the pipe loses too much heat, it freezes. Plumbers are also concerned about how much heat energy enters a plumbing system. Too much heat can cause damage. That is why, for example, hot wastes are allowed to cool before they enter the sanitary system.

> **DID YOU KNOW?**
> **Microwave Ovens—The Power to Move Molecules**
>
> Microwave ovens are a common kitchen appliance. People use microwaves for everything from thawing frozen food to boiling a cup of water and cooking a steak. Microwave ovens cook food faster and more efficiently than conventional ovens. How do they work? Microwaves are tiny radio waves that are absorbed by water molecules, fats, and sugars. Other materials, such as plastic or ceramic, do not absorb microwaves. When microwaves are absorbed, they convert directly into heat.
>
> When a molecule is heated, scientists say it becomes excited. Heated molecules move fast, bounce against one another, and ricochet off container walls. Though scientists can't actually see individual molecules, they can see the effect of their excited state. When heated molecules bump into a larger particle, the molecules cause the particle to move in a zigzag pattern. This form of motion is called Brownian motion, in honor of Robert Brown, a scientist who mathematically described it in 1827.

4.1.1 Thermometers

Thermometers are tools that measure temperature. They are available in various shapes and are seen in numerous applications. Thermometers are installed wherever there is a need to know a temperature range. Always consult the manufacturer's instructions before installing or using a thermometer.

There are three popular types of thermometers. **Liquid thermometers** use a glass tube filled with fluid, such as mercury or alcohol. The fluid expands when heated and rises in the tube. Lines on the tube corresponding to the height of the fluid indicate the temperature. Liquid thermometers are commonly used in refrigerators and freezers (see *Figure 12*). **Bimetallic thermometers** use **thermal expansion** to show a temperature reading. Inside a bimetallic thermometer is a coil of metal. The coil is made of two thinner metal strips which are bonded together. Each of these strips expands at a different rate when heated. This causes the coil to curve when the thermometer is placed near a heat source. An indicator attached to the coil then points to temperature lines marked on a dial (see *Figure 13*). Bimetallic thermometers are found in thermostats and in home and commercial heating/cooling systems. **Electrical thermometers** (sometimes called digital thermometers) convert electrical resistance or voltages generated by heat into a temperature reading. Electrical thermometers usually have a digital readout. Today electrical thermometers with digital displays are installed in some water heaters (see *Figure 14*).

Thermometers display temperatures using the **Fahrenheit scale**, the **Celsius scale**, the **Kelvin scale**, or sometimes all three (see *Figure 15*). The first two scales are widely used in residential, commercial, and industrial thermometers. On the Fahrenheit (F) scale, 32°F is the freezing point of pure water at sea level. The boiling point of pure water at sea level is 212°F.

Figure 12 ♦ Liquid thermometer for refrigerators and freezers.

Figure 13 ♦ Bimetallic thermometers.

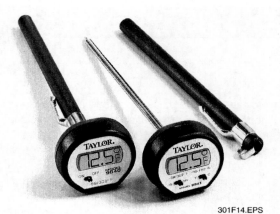

Figure 14 ◆ Electrical thermometers.

Unlike the Fahrenheit scale, the Celsius scale is calibrated to the temperature of pure water at sea level pressure. On the Celsius (C) scale, therefore, 0°C is the freezing point of pure water and 100°C is the boiling point. The Celsius scale is a **centigrade scale** because it is divided into 100 degrees (remember that *centi-* means *one hundredth*).

The Kelvin scale is the basic system of temperature measure in the metric system. Its zero point is the coldest temperature that matter can attain in the universe, according to science. Manufacturers' specifications for fittings and fixtures may be provided in degrees Kelvin.

To convert Fahrenheit measurements into Celsius measurements, first subtract 32 from the Fahrenheit temperature, then multiply the result by ⅝. The formula can be written this way:

C = (F − 32) × 5/9

For example, let's suppose you wanted to find the Celsius equivalent of 140°F. Apply the formula as follows:

C = (140 − 32) × 5/9
C = (108) × 5/9
C = 540/9
C = 60

140°F equals 60°C.

To find the Fahrenheit equivalent of a Celsius temperature, first multiply the Celsius number by ⁹⁄₅. Then add 32. The formula looks like this:

F = (C × 9/5) + 32

To test the formula, practice converting 10°C into Fahrenheit:

F = (10 × 9/5) + 32
F = (90/5) + 32
F = 18 + 32
F = 50

10°C equals 50°F.

Centigrade (Celsius) C°	Fahrenheit F°	Kelvin K°	
100	212	373	WATER BOILS
90	194	363	
80	176	353	
70	158	343	
60	140	333	
50	122	323	
40	104	313	
30	86	303	
20	68	293	
10	50	283	
0	32	273	WATER FREEZES
−10	14	263	
−20	−4	253	

Figure 15 ◆ Celsius, Fahrenheit, and Kelvin temperature scales.

4.1.2 Thermal Expansion

Materials expand when they are heated and contract when they cool. The change in size caused by heat is called thermal expansion. Thermal expansion occurs in all the dimensions of an object: length, width, height, and density. Thermal expansion is affected by the following factors:

- *Length of the material* – the longer the material, the greater the increase in the length.
- *Temperature change* – the greater the temperature, the greater the increase in size.
- *Type of material* – different materials expand at different rates (see *Figure 16*).

Thermal expansion alters the area and volume of the heated material. For pipe, that means diameter, thread size and shape, joints, bends, and length are all affected. These changes could pose a serious problem for the safe and efficient operation of a plumbing system. Always use materials that are appropriate for the temperature range.

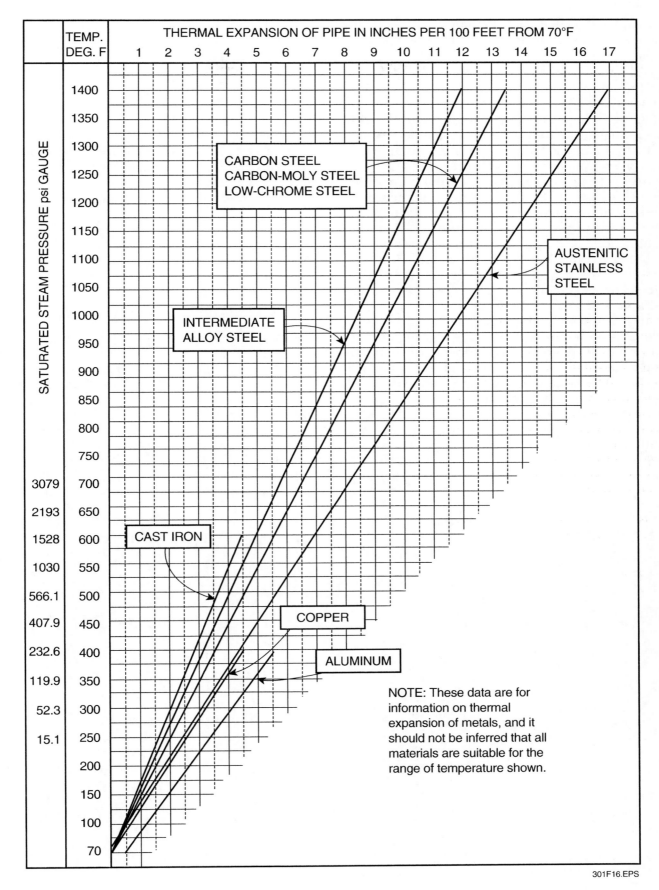

Figure 16 ♦ Thermal expansion of different types of pipe materials.

Consult your local code or talk to a local code expert about temperature considerations in your area.

The rate of thermal expansion of pipe depends on the temperature of the pipe and the type of material the pipe is made from. Refer to the manufacturer's thermal expansion data (see *Table 4*) for information about the type of pipe you are installing. Your local code will provide information on approved methods for controlling thermal expansion, such as the installation of a pressure-reducing valve. The information will also help you install pipe hangers and supports so that they allow for the correct amount of expansion and contraction.

Using the thermal expansion information provided by the manufacturer, you can determine pipe expansion (and contraction) at different temperatures. Consider the following example. Say that you have just installed a 750-foot run of austenitic stainless steel pipe when the air temperature was 50 degrees. Your supervisor instructs you to calculate the length by which the pipe will expand when steam at 400 degrees is allowed to flow through the system.

Begin by finding the column for austenitic stainless steel pipe, and move down the column until you find the expansion coefficient for 400 degrees (3.88). Now find the expansion for 50 degrees (–0.22). Subtract the second expansion coefficient from the first (remember that subtracting negative numbers from positive numbers is the same as adding the numbers):

$x = 3.88 - (-0.22)$
$x = 3.88 + 0.22$
$x = 4.1$

Now multiply the resulting coefficient by the length of the pipe. Notice that the expansion coefficients in *Table 4* are per 100 feet of pipe. This means you must move the decimal point two places to the left for this calculation:

$7.5 \times 4.1 = 30.75$

The pipe length will increase by a total of 30.75 inches over its entire length once it has been completely heated by the steam.

4.1.3 Protecting Pipes From Freezing

Plumbers install pipes, fittings, fixtures, and appliances to minimize repeated expansion and contraction. Underground pipes that are at risk of freezing are installed below the frost line. Depending on where in the country you work, the frost line depth may vary. Consult the local code and building officials. Remember that codes require accessibility to valves on water service pipes and cleanouts on the sanitary sewer drainpipe.

Table 4 Thermal Expansion of Pipe Materials

	Expansion (inches per 100 feet from 70 degrees Fahrenheit)			
Degrees Fahrenheit	Carbon, Carbon /Molybdenum, and Low Chrome Steel	Austenitic Stainless Steel	Copper	PVC
–50	–0.87	–1.23	–1.32	–4.00
0	–0.51	–0.79	–0.79	–2.33
50	–0.15	–0.22	–0.22	–0.67
70	0.00	0.00	0.00	0.00
100	0.23	0.34	0.34	1.00
150	0.61	0.91	0.91	2.67
200	0.99	1.51	1.51	—
250	1.40	2.08	2.08	—
300	1.82	2.67	2.67	—
350	2.26	3.27	3.27	—
400	2.70	3.88	3.88	—
450	3.16	4.49	4.49	—
500	3.62	5.12	5.12	—
550	4.11	5.74	5.74	—
600	4.60	6.39	6.39	—

DID YOU KNOW?
Fahrenheit, Celsius, and Kelvin

The Fahrenheit, Celsius, and Kelvin scales are named after pioneering scientists who helped develop modern methods of temperature measure.

Daniel Gabriel Fahrenheit (1686–1736) was a German physicist. In 1709, he invented a liquid thermometer that used alcohol. Five years later, he invented the mercury thermometer. In 1724, he developed a temperature scale that started at the freezing point of salt water. This is the modern Fahrenheit scale.

Anders Celsius (1701–1744) was a Swedish astronomer and professor who developed a centigrade scale in 1742. He based the scale's zero-degree mark on the freezing point of pure water and the 100-degree mark on the boiling point of pure water. The modern centigrade temperature scale is the same one that Celsius developed, and it is named in his honor.

William Thomson, Lord Kelvin (1824–1907), was a British physicist and professor. He conducted pioneering research in electricity and magnetism. He also participated in a project to lay the first communications cable across the Atlantic in 1857. Kelvin was also an inventor. In the mid-nineteenth century, he developed a measuring system based on the Celsius scale. The zero point of the scale is the theoretical lowest temperature in the universe. The Kelvin scale is the metric system's standard temperature scale.

Snow acts like insulation and keeps frost from penetrating the soil too deeply. Therefore, the frost depth on snow-covered ground is significantly less than it is in other areas that are routinely cleared of snow, such as driveways. Therefore, in colder climates, do not locate the service entrance under a sidewalk, driveway, or patio. Note also that frost penetrates wet soils more deeply than it does dry soils. If possible, locate water pipes under lawns or other areas where snow will remain. Inside buildings, use insulation to protect pipes from freezing.

4.2.0 Pressure and Force

How can a slug of air in a drainpipe blow out a fixture trap seal? Why does water in a tank high above ground flow with more force than water in a tank at ground level? The answer to both questions has to do with pressure. Pressure is the force created by a liquid or gas inside a container. When a liquid or gas is compressed, it exerts more pressure on its container. If the compressed liquid or gas finds an outlet, it will escape. The speed of the escaping liquid or gas is proportional to the amount of pressure to which it was subjected. In the examples above, the pressure applied by the weight of a water column causes both reactions.

In the English system of measurement, pressure is measured in **pounds per square inch** (psi). There are two ways to measure psi:

- *Pounds per square inch absolute (psia)* – uses a perfect vacuum as the zero point. The total pressure that exists in the system is called the absolute pressure.
- *Pounds per square inch gauge (psig)* – uses local air pressure as the zero point. Gauge pressure is pressure that can be measured by gauges.

Psig is a relative measure. Absolute pressure is equal to gauge pressure plus the atmospheric pressure. That's why if you are in Miami (at sea level), 0 psig will be 14.7 psia. However, if you are in Denver (elevation 5,280 feet), 0 psig will be 12.1 psia. The relationship between the two forms of measurement is easy to remember:

psia = psig + local atmospheric pressure

Pressure is related to temperature; increasing the pressure in a container will increase the temperature of the liquid or gas inside. Likewise, when you increase the temperature of a liquid or gas in a closed container, the pressure on the container will rise. This is why steam whistles out of a kettle when you boil water on the stove. If the steam did not escape, the pressure created by the steam would eventually blow off the lid of the kettle. In the case of excess temperature and pressure in a 40-gallon water heater, the consequences would be much more severe.

Pressure is a fundamental concept in plumbing. Two of the most important forms of pressure in plumbing are air pressure and water pressure. Atmospheric air is a gas, and therefore it exerts pressure. Air at sea level (an altitude of zero feet) creates 14.7 pounds of pressure per square inch of surface area. Residential and commercial plumbing installations operate at this pressure. Vents maintain air pressure within a plumbing system. Vents allow outside air to enter and replace wastewater flowing out of various parts of the system.

Water pressure is especially important to plumbers. The next sections discuss water hammer, a damaging effect of pressure, and head, a way of measuring water pressure.

4.2.1 Water Hammer

When liquid flowing through a pipe is suddenly stopped, vibration and pounding noises, called water hammer, result. Water hammer is a destructive form of pressure. The forces generated at the point where the liquid stops act like an explosion. When water hammer occurs, a high-intensity pressure wave travels back through the pipe until it reaches a point of relief. The shock wave pounds back and forth between the point of impact and the point of relief until the energy dissipates (see *Figure 17*). Installing water hammer arresters prevents this type of pressure from reducing the service life of pipes.

The Plumbing & Drainage Institute (PDI) has developed an alphabetical rating system for water hammer arresters. Arresters are rated from A to F by size and shock absorption capability. In the PDI scale, A units are the smallest and F units are the largest. The size of the arrester will depend on the load of the fixtures on the branch. The load factors are represented by a measurement called the **water supply fixture unit (WSFU)**. You will learn how to use WSFUs in the module titled *Sizing Water Supply Piping*. For now, you only need to know that the greater the number of WSFUs on the branch, the larger the water hammer arrester you should install (see *Table 5*).

4.2.2 Head

Water pressure is measured by the force exerted by a water column. A water column can be a well, a run of pipe, or a water tower. The height of a water column is called **head**. Head is measured in feet. The pressure exerted by a rectangular column of water 1 inch long, 1 inch wide, and 12 inches (1 foot) high is 0.433 psig. Another way to express the same measurement is to say that 1 foot of head is equal to 0.433 psig. Each foot of head, regardless of elevation, exerts 0.433 psig. It takes 27.648 inches of head to exert 1 psig.

The pressure, however, is determined solely by vertical height. Fittings, angles, and horizontal runs do not affect the pressure exerted by a column of water. You can use a simple formula to calculate psi:

psi = elevation × 0.433

Use this formula to determine the pressure required to raise water to a specified height and the pressure created by a water column from an elevated supply (see *Figure 18*).

If you want to determine the pressure required to raise water to the top of a 50-foot building, perform the following calculation:

psi = 50 × 0.433
psi = 21.65

ON THE LEVEL

Water Pressure: Pressure Versus Force

If you have two vertical pipes of different diameters, cap the bottom ends of both, and fill them each with water up to the same height (for example, 2 feet of water). The pressure on the cap will be the same in both pipes, but the force of the water will be greater in the wider pipe because there is more water in the pipe.

Pressure is a function of height, not volume, and force is a measure of volume, not height. The reason a higher column of water makes water flow faster through a spigot has nothing to do with the amount of water in the pipe, but has to do with the height of the column.

Think of it this way: take 15 cubic feet of water and put it in a 4-inch pipe. Call the amount of pressure "x" and the amount of force "y." If you took that 15 cubic feet of water and put it in a pipe that was only 2 inches in diameter, the pressure would still be "x" (because the amount of water is the same), but the force would now be "2y" (because the column of water is twice as high).

Plumbers need to know pressure to make sure the fixtures work the way they are designed to and force to make sure the pipes they installed on those fixtures don't break.

Table 5 Sizing a Water Hammer Arrester

PDI Units	A	B	C	D	E	F
WSFUs	1-11	12-32	33-60	61-113	114-154	155-330

To raise a column of water 50 feet, therefore, you need to supply 21.65 psi. If you want to determine the pressure created by a 30-foot head of water in a plumbing stack, you would calculate the following:

psi = 30 × 0.433

You would find that the column of water is creating a force of 12.99 psi on the bottom of the stack.

Consider, for example, a six-floor building with a 150-foot-high tank (refer to *Figure 18*). Note the decrease in pressure as each floor gets higher. You need a force of 43.3 psi on the bottom of the stack, but to raise the column of water to the sixth floor, you need to supply 21.7 psi.

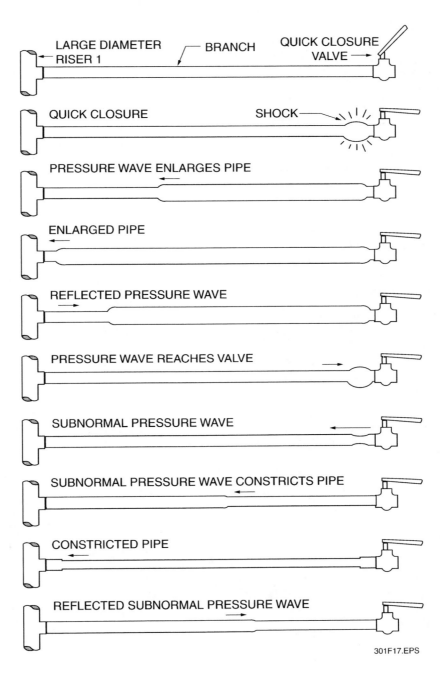

Figure 17 ◆ Water hammer.

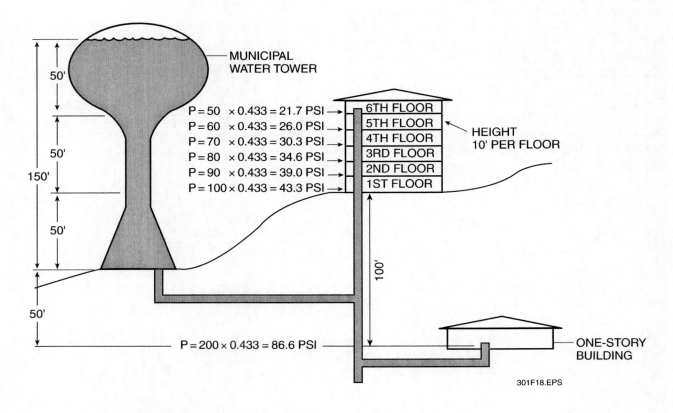

Figure 18 ♦ Water storage tank supplying water to buildings of different heights.

4.2.3 Calculating Force on Test Plugs

Plumbers are responsible for arranging proper testing of drain, waste, and vent (DWV) systems and water supply piping. Water and air tests are the most common ways to ensure compliance with plans and codes. Inspectors block pipe openings with test plugs (see *Figure 19*) and increase the force in the system to test for leaks. You can calculate the force applied to test plugs using the mathematical formulas you have learned in this module.

Calculating the forces on test plugs will help ensure that the plumbing system can withstand operating forces. Test plugs are rated, so for safety reasons it is important to select a test plug that is suited for the pressures and size of the system. Consider the following example. Say you want to determine the total force on a 4-inch-diameter test plug with a water head of 25 feet. Follow these steps:

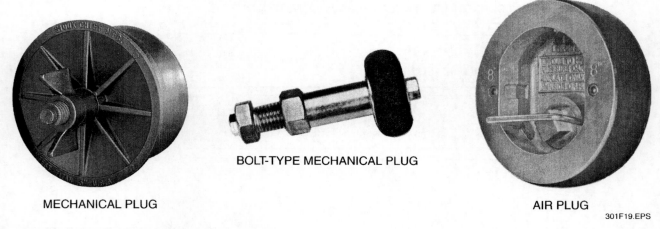

Figure 19 ♦ Examples of air and mechanical test plugs.

Step 1 Calculate the force of the water.
psi = 25 × 0.433
psi = 10.825

Step 2 Calculate the area of the test plug.
A = π × 2²
A = 3.1416 × 4
A = 12.566 sq in

Step 3 Solve for the force on the plug.
psig = (10.825)(12.566)
psig = 136.027

The force on a 4-inch-diameter test plug from a water head of 25 feet is 136.027 psig. The proper head is the pressure required by the local code for a pressure test. The larger the diameter of a vertical pipe, the larger the force exerted on the pipe walls, fixtures, and plugs. Although the pressure is the same as for a narrower pipe of the same height, force varies with the volume.

4.2.4 Temperature and Pressure in Water Heaters

Water under pressure boils at a higher temperature than water at normal pressure. At normal (sea level) pressure, water boils at 212°F. However, at 150 pounds per square inch (psi), water boils at 358°F. You are already familiar with special valves that control temperature and pressure in water heaters (see *Figure 20*). Temperature regulator valves operate when the water gets too hot. Pressure regulator valves operate before the pressure of the water becomes too strong for the water heater to withstand. Combination temperature and pressure (T/P) relief valves prevent damage caused by excess temperatures and pressures. Codes require the installation of T/P relief valves on water heaters. Refer to your local code for specific guidelines. Always follow the manufacturer's instructions to ensure that the heater functions within specified operating temperatures and pressures.

DID YOU KNOW?
The idea of work is common sense, but here is the concept behind it. Work is what happens when force is used to move a load over a distance. The amount of work (W) can be calculated by multiplying force (F) by distance (D): W = F × D.

5.0.0 ◆ SIMPLE MACHINES

Machines and people perform **work**. Whether the work is moving a car on wheels down a road, digging a trench, spinning a drill bit at high speed, or lifting a pallet of materials, machines and people move things by applying some kind of force to them. Muscle power, electricity, air or water pressure, and engines can all supply the force to run machines.

Machines perform work by combining different types of actions or by performing just one action over and over again. A machine that performs a single action is called a **simple machine**.

There are six types of simple machines:

- The **inclined plane**
- The **lever**
- The **pulley**
- The **wedge**
- The **screw**
- The **wheel and axle**

Figure 20 ◆ Relief valve.

DID YOU KNOW?
The Simplicity of Complex Machines

A complex machine is made up of two or more simple machines. Combining different types of simple machines leads to common tools we see everywhere. Wheelbarrows use levers and wheels and axles. Backhoes and cranes both use levers, pulleys, and wheels and axles. Combining tools also allows work to be completed more efficiently. For example, by adding more pulleys, heavier loads can be lifted with less effort. This is called mechanical advantage.

A complex machine is formed when two or more simple machines are combined.

5.1.0 Inclined Planes

An inclined plane is a straight, slanted surface (see *Figure 21*). Loads can be raised or lowered along an inclined plane. Ramps are a common type of inclined plane. The longer the inclined plane is, the easier it is to move a load up or down. In other words, with an inclined plane, greater distances require less force.

A pipe with grade is an inclined plane used regularly in plumbing. You have learned that the steeper the grade, the faster the liquid flows through the angled pipe. Plumbers must install pipe with the correct grade. Too much grade means that liquids flow too fast to scour the pipe clean. Too little grade will allow deposits to settle that will eventually block the pipe.

5.2.0 Levers

A lever is a straight bar, such as a pole or a board, that is free to pivot around a hinge or a support. The hinge or support is called a **fulcrum**. Operate a lever by applying force to the bar. This causes the bar to rise, carrying a load with it. The longer the lever is (that is, the farther the force is from the fulcrum), the easier it is to move the load. There are three types, or classes, of levers (see *Figure 22*):

- *First-class lever* – The fulcrum is located in the middle of the bar. The force is applied in a downward motion to one side of the fulcrum, and the load is on the opposite side of the fulcrum. A playground seesaw is a familiar type of first-class lever.
- *Second-class lever* – The load is between the fulcrum and the force. The force is applied upward. This type of lever requires much less force to move a load. However, the load does not move as far as when a first-class lever is used. Float arms, used on float-controlled valves in toilets, and wheelbarrows are examples of a second-class lever.

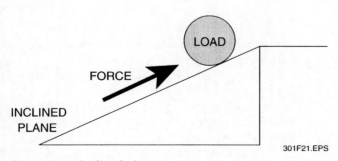

Figure 21 ◆ Inclined plane.

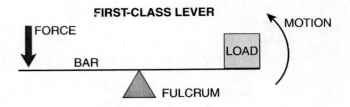
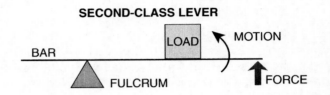
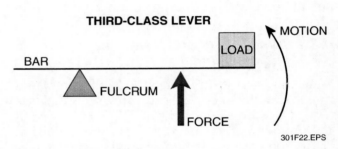

Figure 22 ◆ The three classes of levers.

- *Third-class lever* – The force is applied between the fulcrum and the load. This type of lever requires the most force to move the load. However, the distance traveled by the load is also the greatest of the three types of levers. A leaning ladder that is pushed up into place from underneath is an example of a third-class lever. The ladder is the load, the feet are the fulcrum, and the pushing motion is the force.

A door is a vertical second-class lever in which the fulcrum is the hinge. Many tools are levers. Wrenches and pliers are good examples. Wrenches are second-class levers because the load being moved is close to the fulcrum. In fact, if you look closely, you will see that the load is the fulcrum (see *Figure 23*). Pliers are actually two levers put together. When you use a claw hammer to pry a nail out of wood, you are using a lever. Can you identify what class of lever a claw hammer is?

5.3.0 Pulleys

A pulley is a rope wrapped around a wheel (see *Figure 24*). When the rope is pulled or released, the object being lifted also moves. Pulleys and sets of pulleys both work in two different ways:

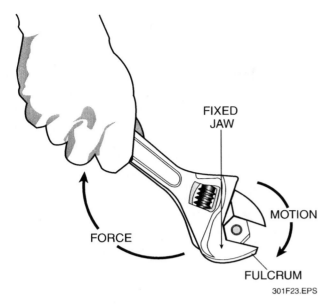

Figure 23 ◆ A wrench is a type of lever.

- They change the direction of the force—for example, pulling down on the rope raises the load.
- They change the amount of force—by combining several pulleys, a lesser force can move a heavier load.

Pulleys are not commonly used on construction projects anymore. However, they can be useful when other options for lifting materials, such as elevators, aren't available. So you should at least be familiar with them in case you need to install one. For example, you could use a chain to turn a gate valve that is high off the ground, thus creating a complex pulley.

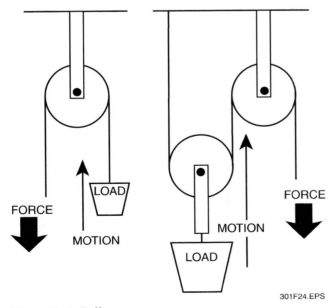

Figure 24 ◆ Pulleys.

5.4.0 Wedges

A wedge is a modified inclined plane. It is made up of two back-to-back inclined planes (see *Figure 25*). Wedges are used to cut or separate objects, prop something up, or hold something in place. You have probably used a wedge to help level a board, hold pipe or batter boards in place, or split wood. A wedge transfers the force along the same line as it is applied. Plumbers use wood chisels and cold chisels, both of which are types of wedges.

5.5.0 Screws

Like the wedge, the screw is a modification of an inclined plane. A screw is really a very long inclined plane that has been wrapped around a cylindrical shaft. When you apply twisting force to a screw perpendicular to the shaft, the plane transforms that into a motion that is parallel to the shaft. Because the inclined plane on a screw is so long, you need less force to move the screw.

Plumbers use screws to fasten pipe clamps and hangers to studs, assemble fixtures and appliance components, and drill wells. Adjustable wrenches use screws to widen and narrow the jaws. Drill bits also use the screw principle (see *Figure 26*).

Figure 25 ◆ Chisels are a type of wedge.

Figure 26 ◆ Drill bits are a type of screw.

5.6.0 Wheels and Axles

The wheel and axle is a modification of the basic pulley concept. The wheel is a circle that is attached to a shaft or post, called an axle, that runs through the center. The wheel and axle turn together (see *Figure 27*). The wheel is probably the most common simple machine. Many tools have wheels and axles. Wheelbarrows use wheels to roll loads along the ground. The handwheel and stem on certain types of valves are another familiar use of the wheel and axle combination.

> **DID YOU KNOW?**
> **The Early Wheel**
>
> The use of wheels without axles dates back to prehistoric times. These wheels were in the form of rollers in which a succession of logs or pipes would be used to carry a load along a road. As the load rolled along, logs would be brought from the back to the front.
>
> The oldest known wheels—from between 3000 to 2000 B.C.E.—were found in Mesopotamian tombs. These were made of wood strips pulled together with cross-struts. A natural knothole acted as the pivot, and the wheel turned on a fixed axle. Wheels with axles were used on Egyptian chariots as early as 1300 B.C.E. and on Roman carts dating from 70 C.E.

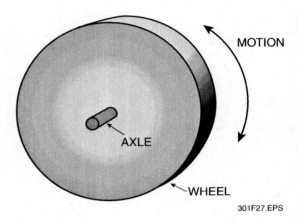

Figure 27 ◆ Wheel and axle.

Review Questions

Sections 4.0.0–5.0.0

1. The _____ scale is the standard metric temperature scale.
   a. Fahrenheit
   b. ampere
   c. mole
   d. Kelvin

2. To convert a Fahrenheit reading to Celsius, you would use the formula _____.
   a. C = (F × %) + 32
   b. C = (F + 32) × %
   c. C = (F – 32) × %
   d. C = (F ÷ 32) × %

3. To find the pressure of a column of water, multiply the height of the column by _____.
   a. 0.433
   b. 3.1416
   c. 14.7
   d. 0.833

4. The force of a water head of 5 feet on a test plug with a diameter of 6 inches is _____.
   a. 41.283 psi
   b. 141.372 psi
   c. 61.214 psi
   d. 244.856 psi

5. In a third-class lever, _____.
   a. the load is located between the fulcrum and the force
   b. the force is applied between the fulcrum and the load
   c. the force is applied outside the fulcrum and load
   d. the fulcrum is located between the force and the load

6.0.0 ♦ THE WORKSHEET

Refer to the appropriate sections in the module to answer the following questions. Remember to show all your work. Round off answers to three decimal places.

1. Convert 3.5 cubic yards to cubic feet.
   _____

2. How many centimeters are in 6.5 dekameters?
   _____

3. How many ounces are in 75 grams?
   _____

4. What is the area of a rectangular roof that is 25 feet 4½ inches by 60 feet 3½ inches?
   _____

5. Refer to *Appendix C*. Determine the area of this shape.
   _____

6. What is the volume of a rectangular prism with a length of 6 feet 3 inches, a width of 2 feet 2¼ inches, and a height of 4 feet?
   _____

7. What is the equivalent temperature in Fahrenheit of 129 degrees Celsius?
   _____

8. What is the pressure of a water head of 75 feet on a test plug with a diameter of 6 inches?
   _____

9. Draw a second-class lever. Show the bar, force, fulcrum, and load.
   _____

10. In what two ways can a pulley work?
    _____

Summary

Applied mathematics—math that you use to accomplish a specific task—is an essential part of plumbing. Plumbers use math in almost every step of the plumbing installation process, from system design and sizing to construction and testing. In this module, you learned about the English and metric systems of measurement. Plumbers use these systems to express how high, how heavy, how much, and how many. You learned how to calculate the area of flat surfaces, such as squares, rectangles, and circles. You also reviewed the formulas for calculating the volume of prisms and columns. You were introduced to the concepts of temperature and pressure and how they affect plumbing installations, and you discovered the mechanical principles behind the tools that you use on the job.

Applied mathematics helps plumbers get a job done correctly. It is as important as any other tool that plumbers use. Take the time to master the basic weights, measures, and formulas. Knowledge of applied mathematics is an essential part of your professional development.

Trade Terms Introduced in This Module

Applied mathematics: Any mathematical process used to accomplish a task.

Area: A measure of a surface, expressed in square units.

Bimetallic thermometer: A thermometer that determines temperature by using the thermal expansion of a coil of metal consisting of two thinner strips of metal.

Celsius scale: A centigrade scale used to measure temperature.

Centigrade scale: A scale divided into 100 degrees. Generally used to refer to the metric scale of temperature measure (see Celsius scale).

Circle: A surface consisting of a curve drawn all the way around a point. The curve keeps the same distance from that point.

Conduction: The transfer of heat energy from a hot object to a cool object.

Cube: A rectangular prism in which the lengths of all sides are equal. When used with another form of measurement, such as cubic meter, the term refers to a measure of volume.

Cubic foot: The basic measure of volume in the English system. There are 7.48 gallons in a cubic foot.

Cubic meter: The basic measure of volume in the metric system. There are 1,000 liters in a cubic meter.

Cylinder: A pipe- or tube-shaped space with a circular cross section.

Decimal of a foot: A decimal fraction where the denominator is either 12 or a power of 12.

Electrical thermometer: A thermometer that measures temperature by converting heat into electrical resistance or voltage.

English system: One of the standard systems of weights and measures. The other system is the metric system.

Equilibrium: A condition in which all objects in a space have an equal temperature.

Fahrenheit scale: The scale of temperature measurement in the English system.

Fulcrum: In a lever, the pivot or hinge on which a bar, a pole, or other flat surface is free to move.

Gallon: In the English system, the basic measure of liquid volume. There are 7.48 gallons in a cubic foot.

Head: The height of a water column, measured in feet. One foot of head is equal to 0.433 pounds per square inch gauge.

Inclined plane: A straight and slanted surface.

Isosceles triangle: A triangle in which two of the sides are of equal length.

Kelvin scale: The scale of temperature measurement in the metric system.

Lever: A simple machine consisting of a bar, a pole, or other flat surface that is free to pivot around a fulcrum.

Liquid thermometer: A thermometer that measures temperature through the expansion of a fluid, such as mercury or alcohol.

Liter: In the metric system, the basic measure of liquid volume. There are 1,000 liters in a cubic meter.

Metric system: A system of measurement in which multiples and fractions of the basic units of measure are expressed as powers of 10. Also called the SI system.

Pounds per square inch: In the English system, the basic measure of pressure. Pounds per square inch (psi) is measured in pounds per square inch absolute (psia) and pounds per square inch gauge (psig).

Pressure: The force applied to the walls of a container by the liquid or gas inside.

Prism: A volume in which two parallel rectangles, squares, or right triangles are connected by rectangles.

Pulley: A simple machine consisting of a rope wrapped around a wheel.

Rectangle: A four-sided surface in which all corners are right angles.

Right triangle: A three-sided surface with one angle that equals 90 degrees.

SI system: An abbreviation for *Système International d'Unités*, or International System of Units. The formal name of the metric system.

Screw: A simple machine consisting of an inclined plane wrapped around a cylinder.

Simple machine: A device that performs work in a single action. There are six types of simple machines: the inclined plane, the lever, the pulley, the wedge, the screw, and the wheel and axle.

Square: A rectangle in which all four sides are equal lengths. When used with another form of measurement, such as square meter, the term refers to a measure of area.

Square foot: In the English system, the basic measure of area. There are 144 square inches in one square foot.

Square meter: In the metric system, the basic measure of area. There are 10,000 square centimeters in a square meter.

Temperature: A measure of relative heat as measured by a scale.

Thermal expansion: The expansion of materials in all three dimensions when heated.

Thermometer: A tool used to measure temperature.

Volume: A measure of a total amount of space, measured in cubic units.

Water Supply Fixture Unit (WSFU): The measure of a fixture's load, depending on water quantity, temperature, and fixture type.

Wedge: A simple machine consisting of two back-to-back inclined planes.

Wheel and axle: A simple machine consisting of a circle attached to a central shaft that spins with the wheel.

Work: A measure of the force required to move an object a specified distance.

Notes

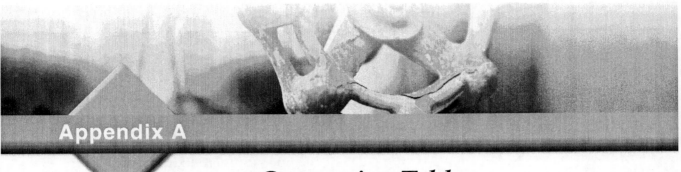

Appendix A

Conversion Tables

A. English to Metric			
	To convert...	**Into...**	**Multiply by...**
LENGTH	Inches	Millimeters	25.4
	Feet	Centimeters	30.48
	Yards	Meters	0.9144
	Miles	Kilometers	1.6
AREA	Square inches	Square centimeters	6.452
	Square feet	Square meters	0.093
	Square yards	Square meters	0.836
	Square miles	Square kilometers	2.589
	Acres	Hectares	0.4047
MASS and WEIGHT	Ounces	Grams	28.35
	Pounds	Kilograms	0.4536
	Short tons	Megagrams (metric tons)	0.907
LIQUID MEASURE	Fluid Ounces	Milliliters	29.57
	Pints	Liters	0.473
	Quarts	Liters	0.946
	Gallons	Liters	3.7854
VOLUME	Cubic inches	Cubic centimeters	16.387
	Cubic feet	Cubic meters	0.0283
PRESSURE	Pounds per square inch	Kilopascals	6.89

B. Metric to English			
	To convert...	Into...	Multiply by...
LENGTH	Millimeters	Inches	0.039
	Centimeters	Feet	0.33
	Meters	Yards	1.0936
	Kilometers	Miles	0.62
AREA	Square centimeters	Square inches	0.155
	Square meters	Square yards	1.196
	Square kilometers	Square miles	0.386
	Hectares	Acres	2.47
MASS and WEIGHT	Grams	Ounces	0.035
	Kilograms	Pounds	2.205
	Megagrams (metric tons)	Short tons	1.1
LIQUID MEASURE	Milliliters	Fluid Ounces	0.0338
	Liters	Pints	2.113
	Liters	Quarts	1.057
	Liters	Gallons	0.264
VOLUME	Cubic centimeters	Cubic inches	0.061
	Cubic decimeters	Cubic feet	0.0353
	Cubic meters	Cubic yards	1.308
PRESSURE	Kilopascals	Pounds per square inch	0.145

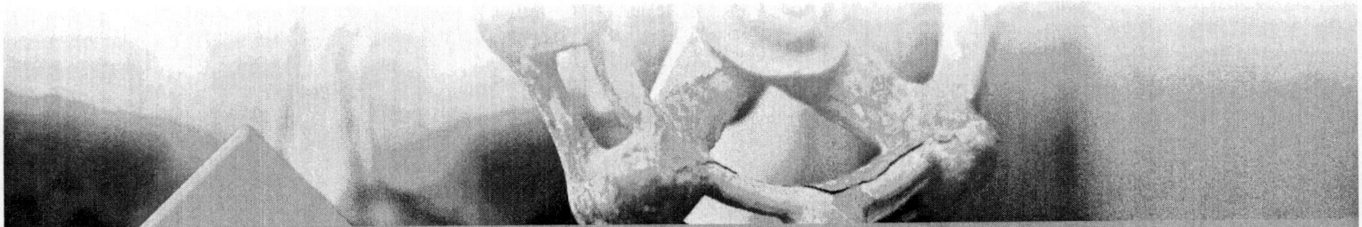

Appendix B

Area and Volume Formulas

Area Formulas

Rectangle:	$A = lw$	(multiply length by width)
Right Triangle:	$A = \frac{1}{2}(bh)$	(one-half the product of the base and height)
Circle:	$A = \pi r^2$	(multiply pi [3.1416] by the square of the radius)

Volume Formulas

Rectangular Prism:	$V = lwh$	(multiply length by width by height)
Right Triangular Prism:	$V = \frac{1}{2}(bwh)$	(multiply one-half the product of base by width by height)
Cylinder:	$V = (\pi r^2)h$	(multiply the area of the circle by the height)

Appendix C

Worksheet Illustration

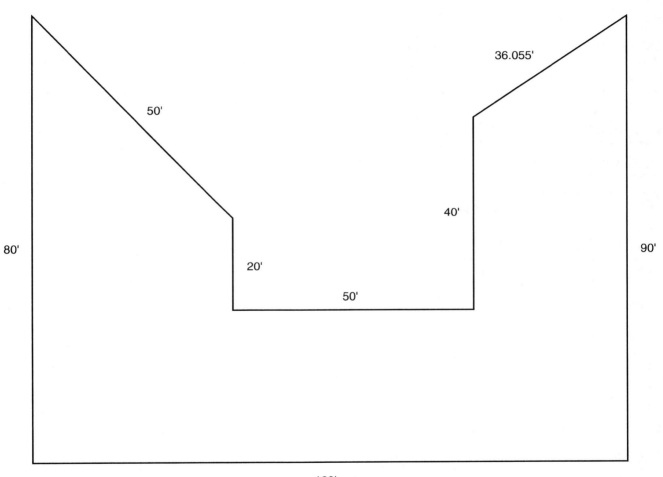

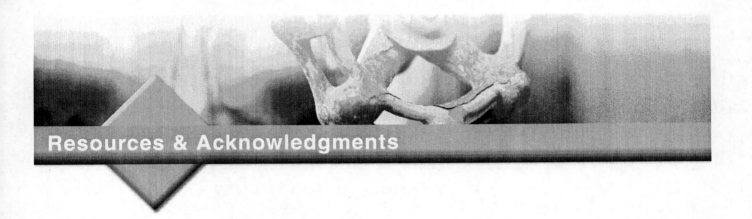

Resources & Acknowledgments

Additional Resources

This module is intended to be a thorough resource for task training. The following reference works are suggested for further study. These are optional materials for continued education rather than for task training.

Code Check Plumbing: A Field Guide to Plumbing, 2000. Redwood Kardon, Michael Casey, and Douglas Hansen. Newtown, CT: Taunton Press.

Explosion Danger Lurks [videotape], ca. 1955. North Andover, MA: Watts Regulator Company.

Math to Build On: A Book for Those Who Build, 1993. Johnny and Margaret Hamilton. Clinton, NC: Construction Trades Press.

References

Code Check Plumbing: A Field Guide to Plumbing, 2000. Michael Casey, Douglas Hansen, and Redwood Kardon. Newton, CT: Taunton Press.

Dictionary of Architecture and Construction, Third Edition, 2000. Cyril M. Harris, ed. New York: McGraw-Hill.

Pipefitters Handbook, 1967. Forrest R. Lindsey. New York: Industrial Press Inc.

Figure Credits

Calculated Industries, Inc., 301F04

Expansion Seal Technologies, 301F19B, 301F19C

Mason Industries, Inc., Table 4

Milwaukee Electric Tool Corporation, 301F26

Plumbing & Drainage Institute, 301F17, Table 5

Ridge Tool Company / Ridgid®, 301F25

Sioux Chief Manufacturing Company, Inc., 301F19A

Taylor Precision Products, 301F12, 301F13, 301F14

Watts Regulator Co., 301F20

CONTREN® LEARNING SERIES — USER UPDATE

The NCCER makes every effort to keep these textbooks up-to-date and free of technical errors. We appreciate your help in this process. If you have an idea for improving this textbook, or if you find an error, a typographical mistake, or an inaccuracy in NCCER's Contren® textbooks, please write us, using this form or a photocopy. Be sure to include the exact module number, page number, a detailed description, and the correction, if applicable. Your input will be brought to the attention of the Technical Review Committee. Thank you for your assistance.

Instructors – If you found that additional materials were necessary in order to teach this module effectively, please let us know so that we may include them in the Equipment/Materials list in the Annotated Instructor's Guide.

Write: Product Development and Revision
National Center for Construction Education and Research
P.O. Box 141104, Gainesville, FL 32614-1104

Fax: 352-334-0932

E-mail: curriculum@nccer.org

Craft _____ Module Name _____

Copyright Date _____ Module Number _____ Page Number(s) _____

Description
_____
_____
_____
_____

(Optional) Correction
_____
_____
_____

(Optional) Your Name and Address
_____
_____
_____

Plumbing Level Three

02302-06

Sizing Water Supply Piping

02302-06
Sizing Water Supply Piping

Topics to be presented in this module include:

1.0.0	Introduction	2.2
2.0.0	Factors Affecting Water Supply Piping	2.2
3.0.0	Laying Out the Water Supply System	2.6
4.0.0	Sizing Water Supply Piping	2.11
5.0.0	The Worksheet	2.15

Overview

The water supply system is one of the most important systems installed in residential and commercial buildings. Plumbers must size supply systems correctly so that they reliably provide adequate water at the correct pressure. Proper installation of water supply systems requires an understanding of the physical properties of water. Each of these factors—temperature, density, flow, and friction—has an effect on the system's operation. Installing a system so that these effects are limited will help the system last longer and operate more efficiently.

Although water supply systems differ from building to building, they are all built using the same basic rules. Working from the material takeoff, plumbers determine the system's water requirements from each fixture's flow rate and flow pressure and the total length and capacity of the pipes and fittings. Then, to estimate the system's demand, plumbers use the load value of each fixture (in WSFUs) to calculate the rate of flow in gpm, also taking into account intermittent or continuous demand and maximum probable flow.

After the system's requirements and demands are estimated, plumbers calculate the correct pipe sizes for the system. There is more than one right way to do this, since plumbers in different areas may use different engineering practices. Local codes provide tables, graphs, and specifications to help with these calculations. Following the highest professional standards in sizing systems helps plumbers to install water supply systems that meet the needs of customers.

Focus Statement
The goal of the plumber is to protect the health, safety, and comfort of the nation job by job.

Code Note
Codes vary among jurisdictions. Because of the variations in code, consult the applicable code whenever regulations are in question. Referring to an incorrect set of codes can cause as much trouble as failing to reference codes altogether. Obtain, review, and familiarize yourself with your local adopted code.

Portions of this publication reproduce tables and figures from the *2003 International Plumbing Code* and the *2000 IPC Commentary*, International Code Council, Inc., Falls Church, Virginia. Reproduced with permission. All rights reserved.

Objectives

When you have completed this module, you will be able to do the following:

1. Calculate pressure drops in a water supply system.
2. Size pipe for different acceptable flow rates.
3. Explain the difference between and advantages of a continuous-flow system and an intermittent-flow system.
4. Identify fixtures with high flow rates.
5. Explain how friction and flow impact a water supply system.
6. Lay out a water supply system.
7. Calculate developed lengths of branches for a given water supply system.
8. Calculate flow rates for high flow rate fixtures.

Trade Terms

Demand
Density
Developed length
Equivalent length
Flow rate
Friction

Friction loss
Laminar flow
Pressure drop
Transient flow
Turbulent flow
Viscosity

Required Trainee Materials

1. Appropriate personal protective equipment
2. Sharpened pencils and paper
3. Copy of local applicable code
4. Calculator

Prerequisites

Before you begin this module, it is recommended that you successfully complete *Core Curriculum; Plumbing Level One; Plumbing Level Two; Plumbing Level Three*, Module 02301-06.

This course map shows all of the modules in the third level of the *Plumbing* curriculum. The suggested training order begins at the bottom and proceeds up. Skill levels increase as you advance on the course map. The local Training Program Sponsor may adjust the training order.

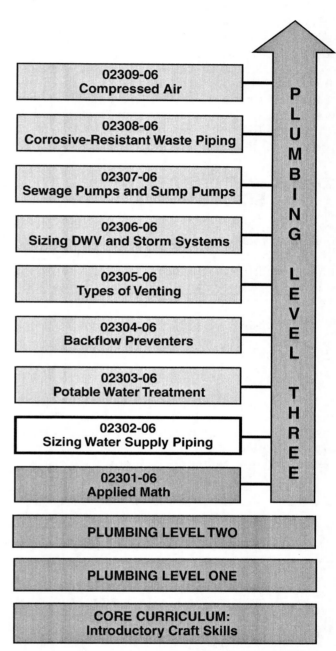

1.0.0 ◆ INTRODUCTION

The water supply system is one of the most important plumbing installations. It is a vital part of residential and commercial buildings. Sizing water supply systems is a complex and important task. Supply systems need to provide adequate water at the correct pressure, and they need to do this reliably and efficiently.

You have already learned how to install water supply piping. This module discusses the general concepts used for sizing water supply systems. You will learn how to calculate a system's requirements and **demand**. When you are finished, you will be able to size a water supply system so that it provides the right amount of water, at the right pressure, at the right time.

2.0.0 ◆ FACTORS AFFECTING WATER SUPPLY PIPING

To install water supply systems correctly, you must understand the physical properties of water. These properties affect how water behaves inside pipes and fittings. Plumbers must consider the following factors when sizing a water supply system:

- Temperature
- **Density**
- Flow
- **Friction**

Each of these factors has an effect on the operation of a water supply system. It is important to install water supply systems so that those effects are controlled. This will help the system last longer and operate more efficiently.

2.1.0 Temperature and Density

You have already learned about the importance of temperature and of one of its related properties, pressure. Another property of temperature that affects plumbing systems is density. Density is the amount of a substance in a given space. It is measured in pounds per cubic foot. Increasing the temperature of a substance decreases its density (see *Figure 1*). It also increases its volume. Plumbers need to know the temperature of water in a plumbing system to choose the right size pipe.

Another way to explain density is to say that as a volume of water gets hotter, it gets lighter. At 32°F, a cubic foot of water weighs 62.42 pounds, but at 100°F, the same amount of water weighs 61.99 pounds. At 200°F, it weighs 60.14 pounds.

Overheated water is dangerous. It can split pipes and weaken fittings. It can also cause water heaters to explode. Install temperature/pressure relief valves on hot water heaters to prevent overheated

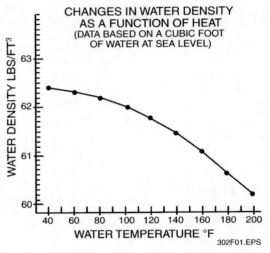

Figure 1 ◆ Density of a cubic foot of water at different temperatures.

DID YOU KNOW?
Facts and Legends About Water

Water is one of the most plentiful substances on Earth. It is common in all living things. It is the internationally recognized standard for temperature and density measurements. However, water behaves very differently from other liquids. Sometimes, water seems to defy common sense. For example, if water begins to freeze, the volume will decrease until it reaches 39°F, then it will start to expand. The same thing happens when you heat frozen water. Water can't be compressed as much as other liquids. Some people even believe that it takes hot water less time to freeze than cold water!

Codes are strict about the sources, quantity, and quality of potable water. They also list the sizes and materials allowed for water supply piping. Plumbers can use only pipe and fittings that are approved by national standards organizations. Violators may face strict penalties. It is in your best interest and that of public safety to follow the local code closely.

water from causing damage (see *Figure 2*). Relief valves release water or steam and equalize the pressure in the tank. Review your local code to determine the proper precautions for hot water pipes.

2.2.0 Flow

In plumbing, flow is the measure of how much liquid moves through a pipe. The measure of how much a liquid resists flow is called its **viscosity**. You may have encountered this term in reference to motor oil. High-viscosity oil is very thick. In other words, it resists flowing. The concepts of flow and resistance are very important in water supply systems. There are three types of flow:

- **Laminar flow**
- **Transient flow**
- **Turbulent flow**

In cases where the rate of flow is less than one foot per second, water flows smoothly. If you could see the molecules in slow-moving water, they would look like they were moving parallel to each other in layers. This is called laminar flow. Laminar flow is also called streamline flow or viscous flow. When water flow changes from one type of flow to another, it becomes erratic and unstable. This type of flow is called transient flow.

Turbulent flow is the random motion of water as it moves along a rough surface, such as the inner surface of a pipe. The faster the water flows along the inner surface, the more turbulent the flow becomes. Turbulent flow is a concern to plumbers, because it can accelerate the wear on pipes.

The different types of flow can be visualized by thinking about how water behaves in a kitchen sink. The water coming out of the faucet moves in a straight line. This is laminar flow. In the basin, it sloshes and spins. This is turbulent flow. The moment at which the water hits the basin and starts to swirl—you may not even be able to see this—is transient flow.

2.3.0 Friction

Turbulent flow is caused by friction. Friction is the resistance or slowing down that happens when two things rub against each other. For example, run your hand over a smooth surface, like silk or a polished tabletop. Your hand moves easily over the surface. It is harder to run your hand over a bumpy rug or over sandpaper because the uneven surface slows your hand down. This is an example of friction.

Friction makes water supply systems less efficient. It reduces the pressure in the system by slowing water down. This is called **friction loss**. Friction loss is also known as pressure loss. In water supply piping, friction is caused by several factors, including:

- The inner surface of a pipe
- Flow through water meters
- Flow through fittings such as valves, elbows, and tees
- Increased water velocity

Effective system installation can reduce overall friction and improve system efficiency. Plumbers can reduce friction by using pipes with a greater diameter. This is especially true when the water pipe is undersized. When water meters are required, use one that is designed for low friction. Keep the number of valves to a minimum to prevent excessive flow restriction. See the section on sizing water supply piping in this module to learn how to calculate friction loss in the system.

There is much less friction at the center of a pipe than near its inner surface. This means that the **flow rate**, or the speed that water flows, is faster at the center of a pipe and slower near the pipe's inner surface (see *Figure 3*). Flow rate is measured in gallons per minute (gpm). Plumbers install plumbing systems to minimize turbulence and maximize flow. Plumbers can consult published sources to determine the average rate of flow under different conditions. Average rate of flow is determined by a combination of available pressure, demand, system size, and pipe size.

Fittings and valves can also cause friction loss. You can calculate the amount of friction loss caused by fittings and valves by comparing them with lengths of pipe that would cause equal friction loss (see *Table 1*). These are called **equivalent lengths**. You can use equivalent lengths to calculate **pressure drop**. This is the difference in pressure between the inlet and the farthest outlet. You will learn more about pressure drop later in this module.

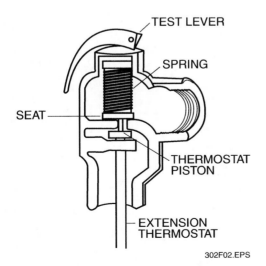

Figure 2 ◆ Temperature/pressure relief valve.

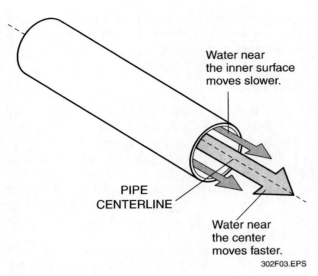

Figure 3 ◆ Pipe flow rates.

Refer to the table to determine the most efficient choice of fittings. For example, say you are going to install two 45-degree elbows to make a right-angle bend. The system uses ⅝-inch tubing. Consult *Table 1*. One 45-degree elbow causes the same friction loss as 0.5 feet of pipe. Therefore, the friction losses in two 45-degree elbows equal the loss in 1 foot of pipe. However, one 90-degree elbow causes the same friction loss as 1.5 feet of pipe. Therefore, if the length of pipe between the two 45-degree elbows was less than 0.5 foot, the use of two 45-degree elbows would be the more efficient choice (see *Figure 4*). Always refer to local code for standards in your area.

The table also allows you to compare the friction losses caused by various types of valves. A rule of thumb that many plumbers use is to allow an additional 50 percent of the system's total length for fittings and valves. You will learn how to calculate the total length later in this module. Remember to test the accuracy of your estimate after the pipe sizes have been determined.

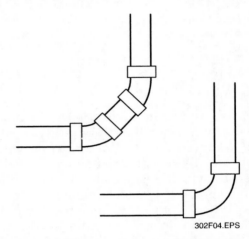

Figure 4 ◆ Friction loss comparison between two 45-degree elbows and one 90-degree elbow.

Table 1 Sample Equivalent Pipe Lengths Used for Determining Friction Loss in Fittings

	Friction Loss in Fittings and Valves Expressed as an Equivalent Length of Tube[a], in Feet								
Nominal or Standard Size (inches)	FITTINGS					VALVES			
	Standard Ell		90-Degree Tee						
	90-Degree	45-Degree	Side Branch	Straight Run	Coupling	Ball	Gate	Butterfly	Check
⅜	0.5	—	1.5	—	—	—	—	—	1.5
½	1.0	0.5	2.0	—	—	—	—	—	2.0
⅝	1.5	0.5	2.0	—	—	—	—	—	2.5
¾	2.0	0.5	3.0	—	—	—	—	—	3.0
1	2.5	1.0	4.5	—	—	0.5	—	—	4.5
1¼	3.0	1.0	5.5	0.5	0.5	0.5	—	—	5.5
1½	4.0	1.5	7.0	0.5	0.5	0.5	—	—	6.5
2	5.5	2.0	9.0	0.5	0.5	0.5	0.5	7.5	9.0
2½	7.0	2.5	12.0	0.5	0.5	—	1.0	10.0	11.5
3	9.0	3.5	15.0	1.0	1.0	—	1.5	15.5	14.5
3½	9.0	3.5	14.0	1.0	1.0	—	2.0	—	12.5
4	12.5	5.0	21.0	1.0	1.0	—	2.0	16.0	18.5
5	16.0	6.0	27.0	1.5	1.5	—	3.0	11.5	23.5
6	19.0	7.0	34.0	2.0	2.0	—	3.5	13.5	26.5
8	29.0	11.0	50.0	3.0	3.0	—	5.0	12.5	39.0

Source: International Code Council, Inc.

Review Questions

Sections 1.0.0–2.0.0

1. When sizing a water supply system, plumbers must consider temperature, density, flow, and _____.
   a. friction
   b. turbulence
   c. pressure
   d. demand

2. Increasing the temperature of a substance _____ its density.
   a. does not alter
   b. equalizes
   c. increases
   d. decreases

3. Density is measured in _____.
   a. pounds per square inch
   b. feet per second
   c. pounds per cubic foot
   d. cubic feet per minute

4. At 100°F, a cubic foot of water weighs _____ pounds.
   a. 64.24
   b. 62.42
   c. 61.99
   d. 60.14

5. Install a _____ on a hot water heater to prevent damage caused by overheated water.
   a. temperature/pressure relief valve
   b. water hammer arrester
   c. check valve
   d. pressure arrester

6. Laminar flow is _____.
   a. Flow occurring as water changes from one flow pattern to another
   b. Smooth flow at or below 1 foot per second
   c. Random flow along a rough surface
   d. Flow resulting from high viscosity fluid moving on a smooth surface

7. Turbulent flow is _____.
   a. Flow resulting from high viscosity fluid moving on a smooth surface
   b. Flow occurring as water changes from one flow pattern to another
   c. Smooth flow at or below 1 foot per second
   d. Random flow along a rough surface

8. _____ flow is caused by friction.
   a. Laminar
   b. Viscous
   c. Turbulent
   d. Transient

9. A viscous liquid tends to flow _____ than a nonviscous liquid.
   a. more slowly
   b. more quickly
   c. more freely
   d. farther

10. Increasing the velocity of water in a pipe _____ friction.
    a. decreases
    b. increases
    c. initiates
    d. does not affect

11. Friction _____ the pressure in a water supply system.
    a. increases
    b. reduces
    c. does not affect
    d. eliminates

12. Each of the following is a common cause of friction in a water supply system *except* _____.
    a. flow through water meters
    b. flow through valves, elbows, and tees
    c. increased water velocity
    d. increased water temperature

Review Questions

13. One way to reduce friction caused by pipes is to use _____ pipe.
    a. smaller diameter
    b. shorter
    c. larger diameter
    d. longer

14. Flow rate is measured in _____.
    a. drain fixture units
    b. gallons per minute
    c. cubic feet per minute
    d. water supply fixture units

15. Refer to *Table 1*. Two 45-degree standard ells are *more* efficient than a single 90-degree standard ell in all pipe size(s) over _____.
    a. ⅜ inch
    b. ½ inch
    c. ⅝ inch
    d. 1 inch

3.0.0 ◆ LAYING OUT THE WATER SUPPLY SYSTEM

Plumbers make sure that water supply systems work. Although systems differ from building to building, they are all built using the same basic rules. Plumbers must be able to determine the water requirements for all of the fixtures and outlets in the system. They must calculate the total length of pipes and fittings and how much water they can handle. All this information helps plumbers install a system that will meet the needs of the building's occupants. In this section, you will learn how to determine system requirements and demand in a water supply system.

3.1.0 System Requirements

Plumbers must consider a number of factors when installing a water supply system. Those factors include the following:

- The system's ability to carry an adequate supply of water.
- The system's ability to operate with a minimum amount of turbulence.
- The system's ability to maintain a reasonable flow rate, which will ensure quiet operation.
- The system's design—it should not be expensively over-designed.

Careful installation will ensure that the system provides the best possible service. Before you begin, consult the building's plans or blueprints. Plumbing drawings are included in the plans for most industrial and commercial buildings. For light residential buildings, consult the floor plan. The drawings will help you understand how the system should be laid out.

Refer to the material takeoff for the types and number of fixtures and outlets used in the system. Determine the flow rate at each fixture and outlet by referring to charts and tables in your local applicable code (see *Table 2*). This information can also be found in the product specifications. Identify both the rate and pressure of flow for each item. Note that the rate and pressure may be less

ON THE LEVEL

Water Pressure Booster Systems

Sometimes the water pressure serving a building is too low for the system's requirements. In these cases, codes require that water pressure booster systems be installed. Fixture outlet pressures vary from code to code. The *International Plumbing Code® (IPC)*, for example, requires a flow pressure of 8 pounds per square inch (psi) for most common household fixtures. A temperature-controlled shower requires up to 20 psi, and some urinals require 15 psi. Check your local code's pressure requirements and the manufacturer's instructions before installing the system.

if water conservation devices are used. Determining the flow rate will provide an initial idea of the size of the system.

One of the goals of an efficient system installation is to supply adequate water pressure to the farthest point of use during peak demand. If the system can do that, then the points in between are also likely to have adequate service. However, this assumption may not always be true. Pressure and water volume requirements will vary from fixture to fixture. Keep these issues in mind when installing the water supply system.

Some fixtures may have a recommended pressure that is lower than the system's pressure. To correct this, install a flow restrictor on the line before the fixture. Flow restrictors can also reduce the system's water and energy consumption. This is especially true for hot water lines. In a high-rise building, pressure at one floor may exceed that at another. Install pressure-reducing valves to balance the pressure throughout the system.

Correct pressure must be maintained within the system. Too little pressure will result in poor service. Negative pressure can cause back siphonage, which could contaminate the system. It can also cause the system to emit an objectionable whistling noise. Whistling can also be caused by pipes that are too small.

ON THE LEVEL

Standards Organizations in the IPC

Plumbers using codes based on the *International Plumbing Code® (IPC)* must use pipes and fittings that conform to strict standards. These standards have been developed by several organizations. Two of the most frequently cited are located in the United States. Another is in Canada.

Take the time to obtain and read these standards carefully. Talk to local plumbing experts and learn how the standards are used in your area. This information will help you broaden your professional knowledge. It will also be reflected in the quality of your work.

American Society for Testing and Materials International (ASTM)
100 Barr Harbor Drive
West Conshohocken, PA 19428-2959
www.astm.org

ASTM develops voluntary consensus standards and related technical information. It provides public health and safety guidelines, reliability standards for materials and services, and standards for commerce.

Canadian Standards Association (CAN/CSA)
5060 Spectrum Way, Mississauga, Ontario
L4W 5N6, Canada
www.csa.ca

CSA develops standards for public safety and health, the environment, and trade. It also provides education and training for people who use their standards.

American Water Works Association (AWWA)
6666 W. Quincy Ave., Denver, CO 80235
www.awwa.org

AWWA is a not-for-profit organization dedicated to improving the quality and supply of drinking water. It specializes in scientific research and educational outreach. AWWA is the world's largest organization of water supply professionals.

Table 2 Demand and Flow Pressure Values for Fixtures

Fixture Supply Outlet Serving	Flow Rate[a] (gpm)	Flow Pressure (psi)
Bathtub	4.00	8
Bidet	2.00	4
Combination fixture	4.00	8
Dishwasher, residential	2.75	8
Drinking fountain	0.75	8
Laundry tray	4.00	8
Lavatory	2.00	8
Shower	3.00	8
Shower, temperature controlled	3.00	20
Sillcock, hose bibb	5.00	8
Sink, residential	2.50	8
Sink, service	3.00	8
Urinal, valve	15.00	10
Water closet, blow out, flushometer valve	35.00	25
Water closet, flushometer tank	1.60	15
Water closet, siphonic, flushometer valve	25.00	15
Water closet, tank, close coupled	3.00	8
Water closet, tank, one piece	6.00	20

[a] See your local code for additional requirements for flow rates and quantities.

Source: International Code Council, Inc.

Some fixtures might require a larger volume of water than others. The installation will have to include a larger pipe to supply that fixture. Remember that water flows more slowly through a wide pipe than a narrow one. Ensure that the rate of flow will not fall below the fixture's requirements.

3.2.0 Calculating Demand

After determining the system's requirements, estimate its demand. The demand is the water requirement for the entire system—pipes, fittings, outlets, and fixtures. Plumbers calculate the rate of flow in gpm according to the number of water supply fixture units (WSFUs) that each fixture requires. WSFUs measure a fixture's load, and vary with the quantity and temperature of the water and with the type of fixture (see *Table 3*). Determine the WSFUs for all fixtures and outlets in the system. Then convert the WSFUs to gpm (see *Table 4*). Add up the results for all fixtures and outlets. This is the total capacity of all the fixtures in the water supply system.

Table 3 Load Values Assigned to Fixtures[a]

Fixture	Occupancy	Type of Supply Control	Load Values in WSFUs		
			Cold Water	Hot Water	Total Load
Bathroom group	Private	Flush tank	2.70	1.50	3.60
Bathroom group	Private	Flush valve	6.00	3.00	8.00
Bathtub	Private	Faucet	1.00	1.00	1.40
Bathtub	Public	Faucet	3.00	3.00	4.00
Bidet	Private	Faucet	1.50	1.50	2.00
Combination fixture	Private	Faucet	2.25	2.25	3.00
Dishwashing machine	Private	Automatic	—	1.40	1.40
Drinking fountain	Offices, etc.	3/8-inch valve	0.25	—	0.25
Kitchen sink	Private	Faucet	1.00	1.00	1.40
Kitchen sink	Hotel, Restaurant	Faucet	3.00	3.00	4.00
Laundry trays (1–3)	Private	Faucet	1.00	1.00	1.40
Lavatory	Private	Faucet	0.50	0.50	0.70
Lavatory	Public	Faucet	1.50	1.50	2.00
Service sink	Offices, etc.	Faucet	2.25	2.25	3.00
Shower head	Public	Mixing valve	3.00	3.00	4.00
Shower stall	Private	Mixing valve	1.00	1.00	1.40
Urinal	Public	1-inch Flush valve	10.00	—	10.00
Urinal	Public	3/4-inch Flush valve	5.00	—	5.00
Urinal	Public	Flush tank	3.00	—	3.00
Washing machine (8 lbs.)	Private	Automatic	1.00	1.00	1.40
Washing machine (8 lbs.)	Public	Automatic	2.25	2.25	3.00
Washing machine (15 lbs.)	Public	Automatic	3.00	3.00	4.00
Water closet	Private	Flush valve	6.00	—	6.00
Water closet	Private	Flush valve	2.20	—	2.20
Water closet	Public	Flush valve	10.00	—	10.00
Water closet	Public	Flush valve	5.00	—	5.00
Water closet	Public or Private	Flushometer tank	2.00	—	2.00

[a] For fixtures not listed, loads should be assumed by comparing the fixture to one listed using water in similar quantities and at similar rates. The assigned loads for fixtures with both hot and cold water supplies are given for separate hot and cold water loads and for total load. The separate hot and cold water loads being three-fourths of the total load for the fixture in each case.

Source: International Code Council, Inc.

Table 4 Table for Estimating Demand

Supply System Predominantly for Flush Tanks			Supply System Predominantly for Flush Valves		
Load	Demand		Load	Demand	
Water Supply Fixture Units	gpm (gallons/min.)	cfm (cubic ft./min.)	Water Supply Fixture Units	gpm (gallons/min.)	cfm (cubic ft./min.)
1	3.0	0.041	—	—	—
2	5.0	0.068	—	—	—
3	6.5	0.869	—	—	—
4	8.0	1.069	—	—	—
5	9.4	1.257	5	15.0	2.005
6	10.7	1.430	6	17.4	2.326
7	11.8	1.577	7	19.8	2.646
8	12.8	1.711	8	22.2	2.968
9	13.7	1.831	9	24.6	3.289
10	14.6	1.962	10	27.0	3.609
11	15.4	2.059	11	27.8	3.716
12	16.0	2.139	12	28.6	3.823
13	16.5	2.206	13	29.4	3.930
14	17.0	2.273	14	30.2	4.037
15	17.5	2.339	15	31.0	4.144
16	18.0	2.906	16	31.8	4.241
17	18.4	2.460	17	32.6	4.358
18	18.8	2.51	18	33.4	4.465
19	19.2	2.56	19	34.2	4.572
20	19.6	2.62	20	35.0	4.679
25	21.5	2.8	25	38.0	5.080
30	23.3	3.115	30	42.0	5.614
35	24.9	3.329	35	44.0	5.882
40	26.3	3.516	40	46.0	6.149
45	27.7	3.703	45	48.0	6.417
50	29.1	3.890	50	50.0	6.684
60	32.0	4.278	60	54.0	7.219
70	35.0	4.679	70	58.0	7.753
80	38.0	5.080	80	61.2	8.181
90	41.0	5.481	90	64.3	8.596
100	43.5	5.815	100	67.5	9.023
120	48.0	6.417	120	73.0	9.759
140	52.5	7.018	140	77.0	10.293
160	57.0	7.620	160	81.0	10.828
180	61.0	8.154	180	85.5	11.430
200	65.0	8.689	200	90.0	12.031
225	70.0	9.358	225	95.5	12.766
250	75.0	10.026	250	101.0	13.502
275	80.0	10.694	275	104.5	13.970
300	85.0	11.363	300	108.0	14.437
400	105.0	14.036	400	127.0	16.977
500	124.0	16.576	500	143.0	19.116
750	170.0	22.726	750	177.0	23.661
1000	208.0	27.805	1000	208.0	27.805
1250	239.0	31.950	1250	239.0	31.950
1500	269.0	35.960	1500	269.0	35.960
1750	297.0	39.703	1750	297.0	39.703
2000	325.0	43.446	2000	325.0	43.446
2500	380.0	50.798	2500	380.0	50.798
3000	433.0	57.883	3000	433.0	57.883
4000	535.0	70.182	4000	525.0	70.182
5000	593.0	79.272	5000	593.0	79.272

Source: International Code Council, Inc.

When sizing residential plumbing systems, keep in mind the greater flow rates for newer high-end baths, whirlpools, and showers; and determine whether a fixture or outlet's demand is intermittent or continuous.

Showers and baths have become high-end consumer products. People can now have shower towers, spas, and whirlpool baths installed in their homes (see *Figure 5*). These fixtures demand higher flow rates than normal showers and baths. Consult your local applicable code because it will specify allowable pressures and flow rates. Codes also specify the proper connection to the supply and recirculation systems. Ensure that the supply system can handle the demand. The product's specifications and installation information will provide the required data.

Next, determine whether the water demand will be intermittent or continuous. Outlets such as lawn faucets, sprinkler systems, and irrigation systems are common continuous-demand items. When they are used, the water flow is constant. The system must be able to supply water to the whole structure even when the continuous outlets are operating. Notice that the fixtures listed above are used only at certain times of the year. Lawn sprinklers and irrigation systems are usually used from spring to early fall. Other continuous-demand systems are needed only under special conditions. Fire-suppression sprinklers are an example. Be sure to size fire sprinklers so that they can provide enough water in an emergency. Check the product's instructions and refer to your local applicable code.

> **WARNING!**
> Always use water supply piping that is made from materials specified in the local code. All water supply piping must be resistant to corrosion. The amount of lead allowed in piping materials is strictly limited. Failure to use approved materials could contaminate the potable water supply. This could cause illness and even death.

Sinks, lavatories, water closets, and similar devices are used for about five minutes or less at a time. They are used only intermittently. Consult the local code for requirements in your area. Combine the amount of flow from continuous- and intermittent-demand fixtures and outlets. Factor this information into the earlier rate-of-flow estimate.

Calculate the system's maximum probable flow. This number is an estimate of peak water demand. Each fixture has flow and demand pressure values assigned to it. You will learn how to calculate flow and demand in the module *Water Pressure Booster and Recirculation Systems*. Consult with local plumbing and code experts to learn about the values used in your area.

Finally, determine the **developed length** of each branch line. The developed length is the total length of piping from the supply to a single fixture. This includes pipes, elbows, valves, tees, and water heaters. Refer back to *Table 1* to determine the equivalent lengths for fittings.

Figure 5 ◆ Luxury shower tower installation.

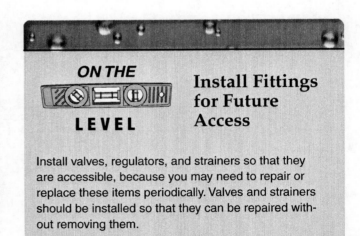

ON THE LEVEL — Install Fittings for Future Access

Install valves, regulators, and strainers so that they are accessible, because you may need to repair or replace these items periodically. Valves and strainers should be installed so that they can be repaired without removing them.

4.0.0 ♦ SIZING WATER SUPPLY PIPING

Once you have estimated the system's requirements and demands, you can begin to size the system. There is more than one right way to size a water supply system. The method discussed in this section is based on the 2003 *International Plumbing Code®* and the 2003 *National Standard Plumbing Code*. Plumbers in your area may use different engineering practices. Take the time to learn how systems are sized in your area. Whatever method you use, ensure that you follow the highest professional standards. The result will be a water supply system that provides the right amount of water at the right amount of pressure.

Begin with an isometric sketch of the entire system (see *Figure 6*). Then, determine the minimum acceptable pressure to be provided to the highest fixture in the system. If adequate pressure is provided to this fixture, you can assume that there will be enough pressure in the rest of the system. Your local code will specify the minimum pressure for flush valves and flush tanks. The system's pressure requirements must not fall below the minimum pressure or exceed the maximum pressure provided by the water supply.

NOTE
Always save copies of the isometric sketches because you will provide them to the owner when you are finished.)

Begin by creating a table, such as the one shown in *Table 5*, to calculate the correct pipe sizes for the system. Divide the isometric drawing of the system

DID YOU KNOW?
History of Pipes and Pipe Sizing in the United States

In the eighteenth and nineteenth centuries, pipe sizing was an art, not a science. Early plumbers used hollowed-out logs. They were limited to the thickness of the logs they could find, generally 9- to 10-inch-wide elm and hemlock trees. Wooden pipes were not ideal because they often sagged, and insects bred in the stagnant puddles that formed at the low points. Plus, log pipes gave water an unappetizing "woody" flavor.

Vents were among the first pipes to be sized. Until the late nineteenth century, vents were designed too small and frequently became clogged with ice or debris. In 1874, an unknown plumber invented a vent that balanced system pressure with the outside air pressure. The system used ½-inch pipe, which was wider than pipes used previously. The plumber also extended the pipe outside the building. It worked for a while, but then plugged up. Through constant experimentation, plumbers eventually discovered the proper size of pipe to use in vents. Plumbers were able to apply this knowledge to sizing supply and drain pipes.

Today's standards for pipe sizing have come a long way since the early days of trial and error.

(refer to *Figure 6*) into sections. The sections should occur where there are branches or changes in elevation. Add each section to the table. If the system design warrants it, consider calculating the hot and cold water requirements separately. Enter the flow in gpm for each section.

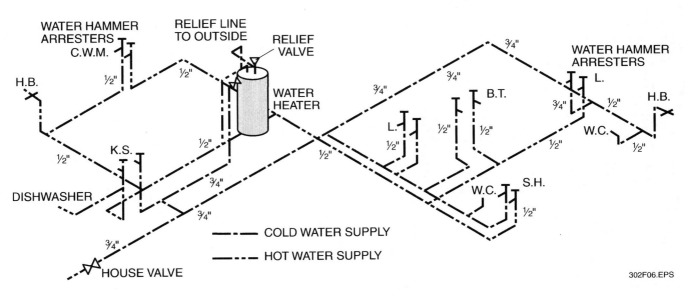

Figure 6 ♦ Water supply system isometric drawing.

Find the minimum pressure available from the supply. Then calculate the highest pressure required in the system. Using your local code, calculate the pressure drop in the system due to the various fittings. Pressure drop is a loss of water pressure caused by:

- Water meters
- Water main taps
- Water filters and softeners
- Backflow preventers
- Pressure regulators
- Valves and fittings (refer to *Table 1*)
- Pipe friction (see *Figure 7*)

Your local code will provide tables and graphs to help you calculate pressure drop. Note that the charts and tables will vary with the pipe material.

Calculate the static head loss, which is the difference in elevation between the supply line and the system's highest fixture. Enter all this information into the table on the appropriate lines and

Table 5 Sample Table for Calculating Pipe Sizes

Col	1		2	3	4	5	6	7	8	9	10
Line	Description		psi or WSFU	gpm through section	Length of section (ft.)	Trial pipe size (in.)	Equivalent length of fittings and valves (ft.)	Total equivalent length, col. 4 + col. 6 (100 ft.)	Friction loss per 100 feet of trial size pipe (psi)	Friction loss in equivalent length, col. 8 × col. 7	Excess pressure over friction loss (psi)
A	Service and cold water distribution piping[a]	Minimum pressure at main	(psi)								
B		Highest pressure required at fixture	"								
C		Meter loss, 2" meter	"								
D		Tap in main loss, 2" tap	"								
E		Static head loss	"								
F		Backflow preventer loss	"								
G		Filter loss	"								
H		Other loss	"								
I		Total losses and requirements	"								
J		Pressure available over pipe friction	"								
K	Pipe section (from diagram)— Cold water distribution piping[b]		(WSFUs)								
	Total pipe friction losses (cold)										
	Difference (line J − line K)										
K	Pipe section (from diagram)— Hot water distribution piping[b]		(WSFUs)								
	Total pipe friction losses (hot)										
	Difference (line J − line K)										

[a] To be considered as pressure gain for fixtures below main (to consider separately omit from "I" and add to "J")
[b] To consider separately, in K use C–F only if greater loss than above.

Source: International Code Council, Inc.

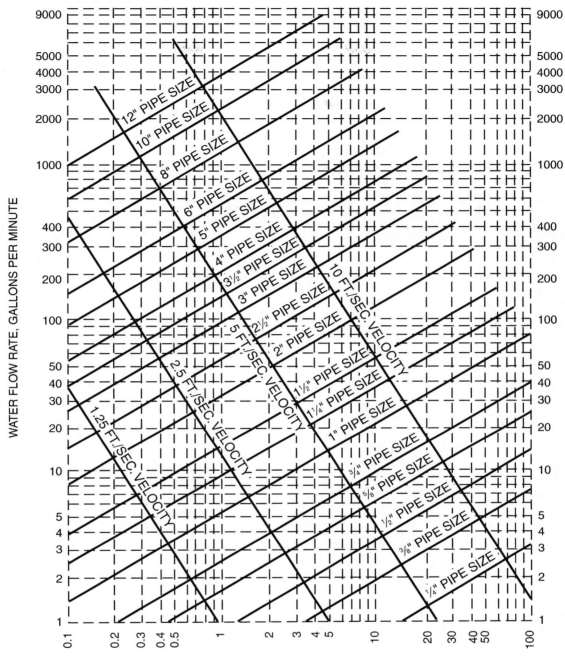

Figure 7 ◆ Sample table for calculating friction loss in smooth pipe[a].

[a] This chart applies to smooth new copper tubing with recessed (Streamline) soldered joints and to the actual sizes of types indicated on the diagram.

add them together. The result is the total overall losses and requirements.

Subtract the total losses from the minimum pressure available. This is the system pressure that is available to overcome friction loss. You can use this information to select the appropriate pipe size for each section of the system. Note that in some systems, the main is above the highest fixture. In this case, the main pressure must be added to the system, instead of subtracted from it.

Add the lengths of all the sections and select a trial pipe size using the following equation:

psi = (available pressure) × 100 ÷ (total pipe length)

Compare this number to the table on friction loss in fittings (refer to *Table 1*). The table indicates the equivalent lengths for trial pipe size of fittings and valves. Add the section lengths and equivalent lengths to find the total equivalent length of the system.

Finally, refer to your local code to determine the friction loss per 100 feet of pipe (refer to *Figure 7*). Multiply this number by the total equivalent length of each section. The result is the friction loss for each section. Add the friction losses together, and subtract the available pressure from the total pipe friction losses. The result is the excess pressure after friction losses. It should be a small positive number. If it is, that means that the trial size pipe is correct for the system. If the result is a large positive number, then the pipe size is too large and can be reduced. Perform a new set of calculations using a new table and the above steps. Always ensure that the final pipe sizes are not less than the minimum size specified in the local code.

Review Questions

Sections 3.0.0–4.0.0

1. Install _____ valves to equalize the pressure throughout a high-rise building's supply system.
   a. balancing
   b. pressure-reducing
   c. check
   d. backflow prevention

2. _____ is the total water requirement for an entire water system, including pipes, fittings, outlets, and fixtures.
   a. Flow
   b. Capacity
   c. Demand
   d. Load

3. Outlets such as lawn faucets, sprinkler systems, and irrigation systems are considered _____ items.
   a. continuous-demand
   b. high-flow
   c. low-use
   d. intermittent-demand

4. The difference in elevation between the supply line and the system's highest fixture is called the _____.
   a. equivalent length
   b. minimum pressure requirement
   c. pressure drop
   d. static head loss

5. To determine the system pressure that is available to overcome friction loss, subtract the total losses from the _____.
   a. total equivalent length
   b. static head loss
   c. minimum pressure available
   d. friction loss

6. Water conservation devices may increase fixture and outlet flow rates and pressures.
   a. True
   b. False

7. Flow pressure is measured in _____.
   a. pounds per square inch
   b. gallons per minute
   c. cubic feet per minute
   d. gallons per second

8. When installing a fixture with a recommended pressure that is lower than system pressure, install a _____ on the line before the fixture.
   a. temperature/pressure relief valve
   b. check valve
   c. flow restrictor
   d. water pressure booster

9. WSFUs vary with each of the following factors *except* _____.
   a. water quantity
   b. water temperature
   c. fixture type
   d. outlet dimensions

10. The total capacity of a water supply system is expressed in _____.
    a. cfm
    b. WSFUs
    c. gpm
    d. psi

Review Questions

11. The last step when calculating demand is to _____.
    a. determine the developed length of each branch line
    b. determine the friction loss on each branch line
    c. calculate the WSFUs for all fixtures and outlets
    d. estimate the system's maximum probable flow

12. When sizing a water supply system, you can assume that there will be enough pressure in the system if adequate pressure is provided to _____.
    a. the lowest fixture in the system
    b. the highest fixture in the system
    c. all continuous-demand fixtures
    d. the fixture with the highest WSFU rating

13. Divide an isometric drawing of a water supply system into sections according to _____.
    a. the number of fixtures or fixture groups
    b. the number of floors in the building
    c. developed length
    d. branches or changes in elevation

14. Static head loss is a measure of differences in _____.
    a. elevation
    b. velocity
    c. pressure
    d. output

15. If a system's excess pressure after friction losses is a _____, then the trial size pipe is correct for the system.
    a. small negative number
    b. large negative number
    c. small positive number
    d. large positive number

5.0.0 ◆ THE WORKSHEET

Use the isometric drawing of a plumbing system in the *Appendix* to answer the following questions. Remember to show all your work.

For Questions 1–6, assume ¾-inch copper piping. Refer to Table 1 in the text or to the appropriate table in your local code.

1. What is the developed length of piping, in feet, for the four fixtures?
   A. _____ B. _____ C. _____ D. _____

2. What is the equivalent length, in feet, for a ¾-inch check valve installed at each of the four fixtures?
   A. _____ B. _____ C. _____ D. _____

3. What is the equivalent length, in feet, for the branch tees in each of the four lines?
   A. _____ B. _____ C. _____ D. _____

4. What is the equivalent length, in feet, for the tee runs in each of the four lines?
   A. _____ B. _____ C. _____ D. _____

5. What is the equivalent length, in feet, for the elbows in each of the four lines?
   A. _____ B. _____ C. _____ D. _____

6. What is the total developed length, in feet, of each of the four lines?
   A. _____ B. _____ C. _____ D. _____

For Questions 7–10, assume a distance of 100 feet from the supply valve. Assume that total friction loss is 10 psi. Refer to *Figure 7* in the text or to the appropriate table in your local code.

7. Find the required pipe size in inches and the velocity in feet per second for Outlet A if it requires 2 gpm.
   Size: _____   Velocity: _____

8. Find the required pipe size in inches and the velocity in feet per second for Outlet B if it requires 4 gpm.
   Size: _____   Velocity: _____

9. Find the required pipe size in inches and the velocity in feet per second for Outlet C if it requires 10 gpm.
   Size: _____   Velocity: _____

10. Find the required pipe size in inches and the velocity in feet per second for Outlet D if it requires 20 gpm.
    Size: _____   Velocity: _____

Summary

A properly sized water supply system ensures that customers have adequate water at the right pressure when they need it. Plumbers install water supply systems tailored for each building. There are several steps and rules that plumbers must follow when installing a supply system. These apply to all systems, no matter how large or small.

To install a system correctly, plumbers must know about the physical properties of water. These properties include temperature, density, flow, and friction. Before installing the system, plumbers also must find the water requirements for all fixtures and outlets. Then they can estimate the total length of all pipes and fittings. They also calculate how much water the system can handle.

All of this information needs to be collected before the pipes can be sized. Pipe sizing involves several steps. The first is to identify the WSFUs for each branch in the system. Next, the plumber selects the right size of pipe for the branches. The plumber then calculates friction loss in the whole system. With that information, the plumber can select the correctly sized supply line. By carefully following all of these steps, the plumber will be able to install an efficient water supply system.

Notes

Trade Terms Introduced in This Module

Demand: The measure of the water requirement for the entire water supply system.

Density: The amount of a liquid, gas, or solid in a space, measured in pounds per cubic foot.

Developed length: The length of all piping and fittings from the water supply to a fixture.

Equivalent length: The length of pipe required to create the same amount of friction as a given fitting.

Flow rate: The rate of water flow in gallons per minute that a fixture uses when operating.

Friction: The resistance that results from objects rubbing against one another.

Friction loss: The partial loss of system pressure due to friction. Friction loss is also called pressure loss.

Laminar flow: The parallel flow pattern of a liquid that is flowing slowly. Also called streamline flow or viscous flow.

Pressure drop: In a water supply system, the difference between the pressure at the inlet and the pressure at the farthest outlet.

Transient flow: The erratic flow pattern that occurs when a liquid's flow changes from a laminar flow to a turbulent flow pattern.

Turbulent flow: The random flow pattern of fast-moving water or water moving along a rough surface.

Viscosity: The measure of a liquid's resistance to flow.

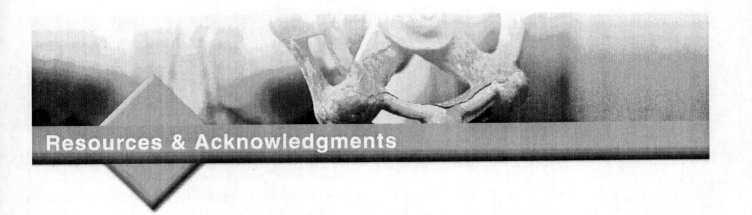

Resources & Acknowledgments

Additional Resources

Code Check Plumbing: A Field Guide to Plumbing, 2000. Redwood Kardon, Michael Casey, and Douglas Hansen. Newtown, CT: Taunton Press.

Plumbers and Pipefitters Handbook, 1996. William J. Hornung. Englewood Cliffs, NJ: Prentice Hall College Division.

Standard Plumbing Engineering Design, Second Edition, 1982. Louis S. Neilsen. New York, NY: McGraw-Hill.

References

Dictionary of Architecture and Construction, Third Edition. 2000. Cyril M. Harris, ed. New York: McGraw-Hill.

Pipefitters Handbook. 1967. Forrest R. Lindsey. New York: Industrial Press Inc.

Figure Credits

International Code Council, Inc.,
2003 International Plumbing Code, 302F07,
Tables 1 through 5

Kohler Co., 302F05

Appendix

Worksheet Illustration

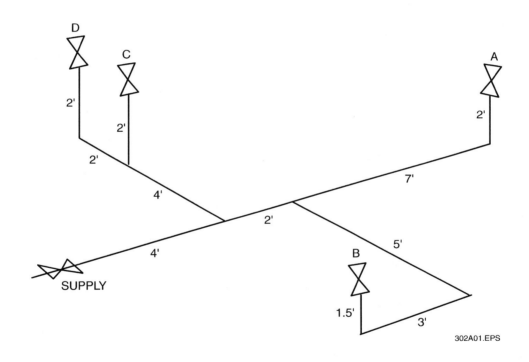

MODULE 02302-06 ◆ SIZING WATER SUPPLY PIPING 2.19

CONTREN® LEARNING SERIES — USER UPDATE

The NCCER makes every effort to keep these textbooks up-to-date and free of technical errors. We appreciate your help in this process. If you have an idea for improving this textbook, or if you find an error, a typographical mistake, or an inaccuracy in NCCER's Contren® textbooks, please write us, using this form or a photocopy. Be sure to include the exact module number, page number, a detailed description, and the correction, if applicable. Your input will be brought to the attention of the Technical Review Committee. Thank you for your assistance.

Instructors – If you found that additional materials were necessary in order to teach this module effectively, please let us know so that we may include them in the Equipment/Materials list in the Annotated Instructor's Guide.

Write: Product Development and Revision
National Center for Construction Education and Research
P.O. Box 141104, Gainesville, FL 32614-1104

Fax: 352-334-0932

E-mail: curriculum@nccer.org

Craft _____ Module Name _____

Copyright Date _____ Module Number _____ Page Number(s) _____

Description
_____
_____
_____
_____

(Optional) Correction
_____
_____
_____

(Optional) Your Name and Address
_____
_____
_____

Plumbing Level Three

02303-06
Potable Water Treatment

02303-06
Potable Water Treatment

Topics to be presented in this module include:

1.0.0 Introduction 3.2
2.0.0 Disinfecting the Water Supply 3.2
3.0.0 Filtering and Softening the Water Supply 3.9
4.0.0 Troubleshooting Water Supply Problems 3.17

Overview

Water is made safe through disinfection, filtration, and softening. While municipal water utilities disinfect their public water supply systems, private water supply systems require their own disinfection devices. The plumber's responsibility for installing water conditioners will vary depending on the local code requirements.

Unfiltered water can reduce the efficiency of a water supply system, shorten the life of appliances, and cause health hazards. Water filters and water softeners separate out undesirable materials. Water softeners use chemicals to remove mineral salts from hard water using a process called ion exchange. One of the most common forms of ion exchange is the zeolite system. Water filters pass water through a porous material such as a membrane or charcoal. Activated carbon (AC) filters are very common and highly effective. Other common types of filters are mechanical, neutralizing, and oxidizing filters. Some water quality problems require extra steps in the filtration process to remove contaminants. In these cases, precipitation systems, reverse osmosis systems, and distillation systems can be used. Plumbers should always test the water to determine the type of contamination, then select the right filtration system.

The most commonly encountered problems in water supply systems are hardness, discoloration, acidity, bad odors and/or flavors, and turbidity. Plumbers must know how to identify the causes and symptoms of the different problems and how to test the water to confirm the diagnosis using the appropriate measurements and treatments. Plumbers should always refer to the local code before attempting to correct a water treatment problem.

Focus Statement
The goal of the plumber is to protect the health, safety, and comfort of the nation job by job.

Code Note
Codes vary among jurisdictions. Because of the variations in code, consult the applicable code whenever regulations are in question. Referring to an incorrect set of codes can cause as much trouble as failing to reference codes altogether. Obtain, review, and familiarize yourself with your local adopted code.

Portions of this publication reproduce tables and figures from the *2003 International Plumbing Code* and the *2000 IPC Commentary*, International Code Council, Inc., Falls Church, Virginia. Reproduced with permission. All rights reserved.

Objectives

When you have completed this module, you will be able to do the following:

1. Flush out visible contaminants from plumbing systems.
2. Disinfect a potable water plumbing system.
3. Identify common water problems and identify the basic equipment to solve them.
4. Practice methods used to soften water.
5. Analyze and measure water-conditioning problems.
6. Install water-conditioning equipment.

Trade Terms

Adsorption
Alum
Backwashing
Calcium hypochlorite
Chlorination
Chlorinator
Coagulation
Contamination
Diaphragm pump chlorinator
Diatomaceous earth
Disinfection
Distillation
Floc
Grains per gallon
Injector chlorinator
Ion
Ion exchange
Osmosis
Oxidizing agent
Pasteurization
Point-of-entry unit
Point-of-use unit
Pollution
Precipitate
Precipitation
Sodium hypochlorite
Solenoid valve
Stroke
Tablet chlorinator
Turbidity unit
Ultraviolet light
Water conditioner

Required Trainee Materials

1. Appropriate personal protective equipment
2. Sharpened pencils and paper
3. Copy of local applicable code
4. Calculator

Prerequisites

Before you begin this module, it is recommended that you successfully complete *Core Curriculum; Plumbing Level One; Plumbing Level Two; Plumbing Level Three,* Modules 02301-06 and 02302-06.

This course map shows all of the modules in the third level of the *Plumbing* curriculum. The suggested training order begins at the bottom and proceeds up. Skill levels increase as you advance on the course map. The local Training Program Sponsor may adjust the training order.

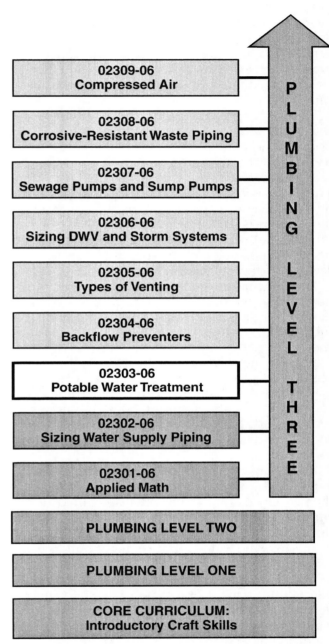

1.0.0 ♦ INTRODUCTION

Until the middle of the nineteenth century, people in Europe and the United States regularly got sick from drinking water. Rural rivers and streams contained sand, clay, and other particles. City water contained germs and discarded chemicals. People who drank such water risked illness and even death. Contaminated water even caused several devastating plagues. The lack of clean water is still a major concern in areas of the world with high poverty, natural disasters, and war. Without clean water, the quality and span of life are drastically reduced.

One important distinction to remember is the difference between **pollution** and **contamination**. Pollution is an impairment in potable water quality that adversely and unreasonably affects the water's aesthetic qualities (making the water unsightly or murky, and/or with an unpleasant odor), but is not hazardous to public health. Contamination is an impairment in potable water quality that creates an actual hazard to public health via poisoning or the spread of disease from sewage, industrial fluids, or waste.

Water is made safe through **disinfection**, filtration, and softening. To disinfect means to destroy harmful organisms in the water. Filtration is the process of scouring water to remove particles and chemicals. Softening, as you have already learned, removes magnesium and calcium salts that cause scale on the inside of pipes and fittings. Water utility companies disinfect, filter, and sometimes soften public water supplies. Some public water requires additional treatment. All well water should be disinfected, filtered, and, if necessary, softened.

In this module, you will learn about the different types of water problems and how to treat them. Devices that treat water are called **water conditioners**. Plumbers can install water conditioners in two locations. Conditioners located near where the water supply enters a building are called **point-of-entry (POE) units**. Conditioners installed at individual fixtures are called **point-of-use (POU) units**. The type of unit to install depends on the water problem and the amount of water to be treated. Often, more than one type of conditioning is required.

The plumber's responsibility for installing water conditioners will vary depending on the local code requirements. Review your local code before you recommend or install a water conditioner. Customers have a right to expect clean, healthful water. It is your responsibility to ensure that their needs are met.

2.0.0 ♦ DISINFECTING THE WATER SUPPLY

If you could look at a drop of untreated water under a microscope, you would see hundreds of tiny living creatures. Most of these creatures are harmless to humans. Many others can cause disease and illness if ingested or inhaled. Certain types of bacteria cause cholera, dysentery, and typhoid fever. Viruses can cause polio, hepatitis, and meningitis. These are just some of the dangers posed by harmful organisms in water. Plumbers prevent contamination of a water supply system due to harmful organisms by disinfecting the system.

You can take steps to prevent contamination while installing a water supply system. Ensure that pipes are stored and handled properly. Do not store pipes in dirty or wet locations. Such places are ideal breeding grounds for harmful organisms. Cap all installed pipe at the end of each workday. When the water supply system has been completely installed, disinfect the system using the following procedure:

Step 1 Flush the completed pipe system with potable water until no more dirty water can be seen coming from any outlet.

Step 2 Fill the entire system with a water/chlorine solution. Valve off the system. Use a solution of at least 50 parts per million (ppm) of chlorine and let the system stand for 24 hours. Alternatively, use a solution of at least 200 ppm of chlorine and let the system stand for 3 hours. Refer to your local code.

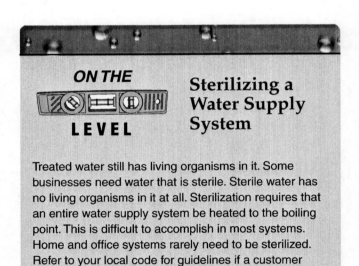

ON THE LEVEL

Sterilizing a Water Supply System

Treated water still has living organisms in it. Some businesses need water that is sterile. Sterile water has no living organisms in it at all. Sterilization requires that an entire water supply system be heated to the boiling point. This is difficult to accomplish in most systems. Home and office systems rarely need to be sterilized. Refer to your local code for guidelines if a customer requires a sterilized water supply system.

Step 3 Flush the system with potable water until the system is completely purged of chlorine. Test the water to determine whether the system is free of chlorine.

Step 4 Have a testing lab or health department test the system for harmful bacteria. If bacteria remain, repeat the previous steps until the system is free of the bacteria.

Municipal water utilities disinfect their public water supply systems. Private water supply systems require their own disinfection devices. Plumbers can choose one of several methods to disinfect a private system. The most common methods are **chlorination, pasteurization**, and **ultraviolet light**. These methods are discussed in more detail below. Take the time to review each method carefully. Discuss them with an experienced plumber. Review your local code to learn when to use these methods in your area.

2.1.0 Chlorination

People who own swimming pools know how important it is to add chlorine to the water regularly. Chlorine kills harmful organisms in the water. The process of using chlorine to disinfect water is called chlorination. Even a small amount of chlorine makes a good disinfectant. A gallon of chlorine bleach at a solution of only 1 ppm will treat about 10,000 gallons of water. The same amount of bleach at a 5 ppm solution will treat 50,000 gallons. Chlorine kills bacteria and viruses within 30 minutes and protects the system for hours after treatment. It also eliminates many of the causes of odor and foul taste. Chlorine can be used in any size water system.

Chlorine is available in liquid and solid forms. The liquid form is called **sodium hypochlorite**. It is commonly found in laundry bleach. A gallon of domestic laundry bleach contains 5.25 percent sodium hypochlorite. Commercial laundry bleach has 19 percent sodium hypochlorite. Because of the higher concentration, commercial bleach is more economical to use for larger water supply systems. The solid form of chlorine is called **calcium hypochlorite**. It is available in powder and tablet form and contains anywhere from 30 percent to 75 percent chlorine.

Chlorine removes iron and sulfur from the water. The process uses large amounts of chlorine. If a system has iron- or sulfur-rich water, monitor and adjust the system's chlorine use regularly. High levels of alkaline in the water may slow the disinfection process. Cold water will also slow the disinfection process. Suspended particles in the water can shield some bacteria from the chlorine.

ON THE LEVEL

Bottled Water—What Are You Drinking?

The bottled water industry has grown rapidly in the past decade. People can buy bottled water in grocery stores and convenience stores. Companies lease and sell water dispensers to homes and offices. Many people now prefer bottled water to tap water. However, in many cases, the source of the water is the same for both. Bottlers are not required to identify where the water comes from.

The U.S. Food and Drug Administration (FDA) regulates bottled water. The FDA has instituted fair labeling rules for bottled water. It has also developed standards for terms such as mineral water, artesian water, and distilled water. Water samples must be tested regularly. All these actions help protect consumers against harm.

Install charcoal filters to remove any residual chlorine taste from the water. Calcium hypochlorite may form deposits in pipes. Always review the manufacturer's instructions before you use any form of chlorine.

Chlorine is distributed through the system by a **chlorinator**. The three most common types of chlorinators are:

- **Diaphragm pump chlorinators**
- **Injector chlorinators**
- **Tablet chlorinators**

The following sections describe each type in more detail.

2.1.1 Diaphragm Pump Chlorinators

Diaphragm pump chlorinators use a diaphragm pump to introduce chlorine into a pressurized water supply system. Diaphragm pumps operate by drawing liquids in and forcing them out in a cycle (see *Figure 1*). Each cycle is called a **stroke**. The diaphragm pump draws chlorine from a storage tank by drawing back the diaphragm. As the diaphragm is pushed out, the chlorine is pumped into a pressurized storage tank. After each cycle, the chlorine mixes with the potable water. The diaphragm pump is set to switch on when the supply system's pump operates. Adjust the amount of chlorine by changing the length of the stroke, the pump's speed, or the amount of time the pump is allowed to operate. A typical diaphragm pump chlorinator installation is illustrated in *Figure 2*.

2.1.2 Injector Chlorinators

Injector chlorinators, also known as aspirators or jets, use a water pump to draw chlorine into the system (see *Figure 3*). The principle is similar to that of a carburetor in a car engine. When the pump turns on, it creates low pressure in the suction line. The low pressure draws water from the pressure tank, where it flows through a device called an injector. As the water flows through the injector, the flow siphons chlorine from a storage tank. The chlorine and water are mixed in the injector. Then the chlorinated water is pumped into the supply system. The amount of chlorine can be adjusted using a setting screw in the injector.

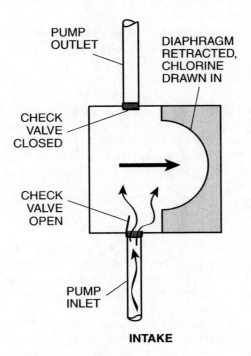
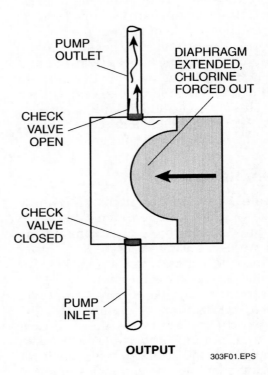

Figure 1 ◆ How a diaphragm pump works.

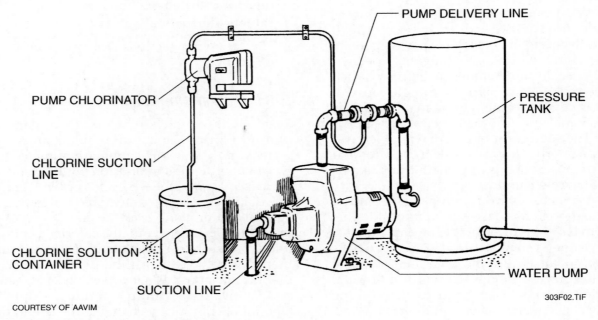

Figure 2 ◆ Typical diaphragm pump chlorinator installation.

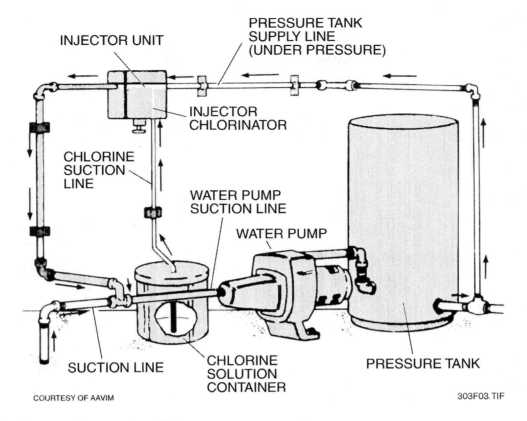

Figure 3 ◆ Typical injector chlorinator installation.

2.1.3 Tablet Chlorinators

In a tablet chlorinator system, water is pumped through a container filled with calcium hypochlorite tablets (see *Figure 4*). A restricting valve on the pump outlet line creates a slight pressure increase on the pump side of the valve and a slight decrease in pressure on the tank side of the valve. The pressure increase diverts some water into the tablet tank, where it is chlorinated. The low pressure then draws the chlorinated water into the pressure tank. The chlorine is diluted in the pressure tank, resulting in the proper solution of chlorine in the water. The amount of chlorine can be adjusted by turning the restricting valve.

 WARNING!
Prolonged exposure to chlorine can cause injury and death. Chlorine can irritate the eyes, nose, and throat. Chlorine vapors can cause lung congestion and emphysema. Liquid chlorine can cause severe burns. Always use appropriate personal protective equipment and follow the manufacturer's instructions when using chlorine.

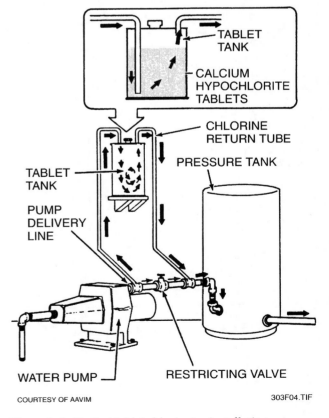

Figure 4 ◆ Typical tablet chlorinator installation.

2.2.0 Pasteurization

The next time you are in the supermarket, look at the label on a milk jug or carton. It may state that the milk is pasteurized. This means that it was heated to kill harmful organisms. Pasteurization can be used to disinfect water, too (see *Figure 5*). Untreated water passes through a heat exchanger, where it is heated to about 150°F. A pump draws the water into an electric heating chamber. There, the temperature of the water is raised to 161°F for at least 15 seconds. This exposure is enough to kill organisms in the water.

After heating, the water flows past a thermostat connected to a **solenoid valve**. A solenoid valve is an electronically operated plunger. If the water is below 161°F, the valve routes the water back to the heating chamber. If the water is still at 161°F, it is sent to the heat exchanger. There, it heats more incoming untreated water. This method can heat only a small amount of water at a time. Treated water is stored in a pressure tank and pumped to the supply system by another pump.

Pasteurization systems are not suitable for all applications. Pasteurization kills harmful organisms much faster than chlorine does and kills both bacteria and viruses. The process is not affected by alkaline levels and does not affect the water's taste. However, most systems can disinfect only 20 to 24 gallons per hour. The heat may cause minerals in the water to settle on the pipe walls as scale. Once water enters the pressure tank, it is not protected from further contamination. Take the time to review the customer's needs before you install this type of system.

> **DID YOU KNOW?**
> ### Louis Pasteur (1822–1895)
>
> In 1864, Louis Pasteur, a French biologist, discovered that heat kills organisms in water. This is one of his most famous discoveries. The process is even named after him: pasteurization. Pasteur also made many other contributions to science. He championed clean and sanitary hospitals as a way to curb the spread of germs. He discovered how to prevent disease through vaccination. His research into the cause of rabies revealed the existence of viruses. For his work, he became world famous. He summed up the secret of his success: "Fortune favors only the prepared mind."

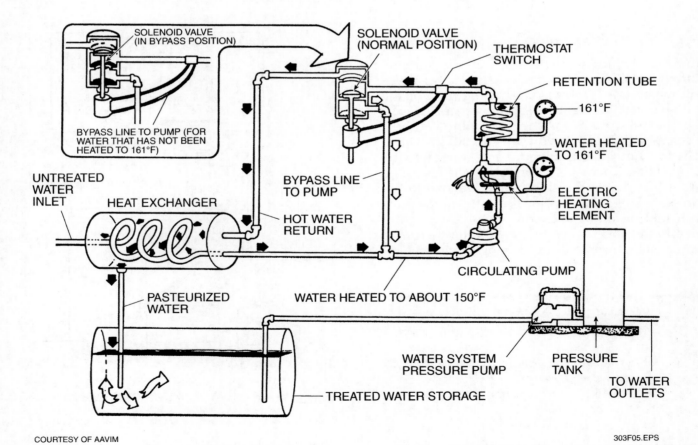

Figure 5 ◆ Typical pasteurization system installation.

2.3.0 Ultraviolet Light

Natural sunlight exposes harmful organisms in streams and rivers to ultraviolet light. This kills the organisms the same way that heat does. Ultraviolet light is a higher frequency than humans can see. In an ultraviolet light system, a solenoid valve directs water from a storage tank into a disinfecting chamber (see *Figure 6*). Inside the chamber is an ultraviolet lamp surrounded by a quartz sleeve. The sleeve prevents the water from cooling the lamp. The ultraviolet light heats the water as it flows through the chamber. The heat kills all bacteria and many viruses. Then the disinfected water flows to the outlet. Ultraviolet light units are available in a range of sizes. Small units can disinfect 20 gallons per hour. The largest units can disinfect up to 20,000 gallons per hour.

Install ultraviolet units on the outlet side of the storage tank. They can also be installed near the fixture outlet. Most ultraviolet units are equipped with a photoelectric cell. This device, also called an electric eye, measures the light given off by the lamp. When an old lamp begins to fade, the eye shuts off the water supply. The old lamp must then be replaced. Install a time delay switch to allow the lamp to reach full intensity before water flows into the chamber. In systems without electric eyes or time delay switches, replace lamps regularly according to the manufacturer's instructions and your local code.

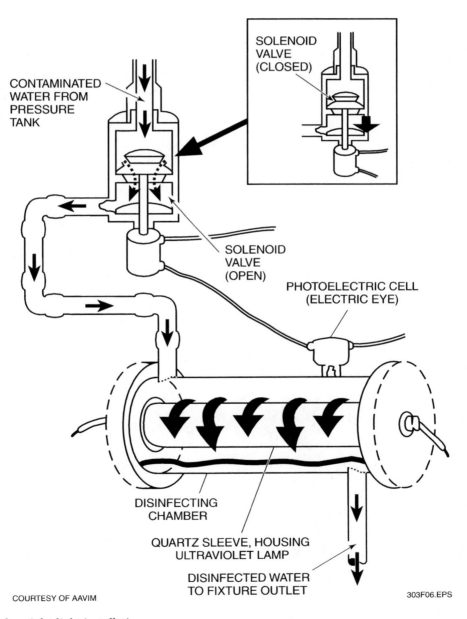

Figure 6 ♦ Typical ultraviolet light installation.

Like pasteurization, ultraviolet light kills organisms quickly and does not affect taste. Suspended particles make the lamp less efficient. Minerals and alkalines can coat the lamp sleeve. Cold water also reduces the lamp's efficiency.

Ensure that filters are installed in the water supply system and that the water temperature is within the correct range, in accordance with the manufacturer's specifications.

ON THE LEVEL

Maintaining Treatment Devices

Ensure that water conditioners are well maintained. Otherwise, pollutants will collect in the device and the results could be worse than the original problem. Install only certified water conditioners. Check the label on the unit. It should say that the unit meets the standards of the National Sanitation Foundation (NSF) or the Water Quality Association (WQA). If it doesn't say that, don't let your customers use it. Don't let your customers risk their health just to save a few dollars.

Review Questions

Sections 1.0.0–2.0.0

1. Each of the following processes is used to make water safe *except* _____.
   a. filtration
   b. softening
   c. disinfection
   d. irradiation

2. _____ removes the chemical salts that cause scale from water.
   a. Filtration
   b. Softening
   c. Chlorination
   d. Disinfection

3. When disinfecting a water supply system, the first step is to _____.
   a. flush the system to remove chlorine
   b. fill the system with a water/chlorine solution
   c. flush the system to remove dirty water
   d. have an inspector test the system for harmful bacteria

4. To kill harmful organisms in a domestic water supply system, fill the entire system with a solution of at least _____ ppm of chlorine and let the system stand for _____ hours.
   a. 25; 3
   b. 30; 12
   c. 50; 24
   d. 200; 12

5. The liquid form of chlorine found in bleach is called _____.
   a. sodium hypochlorite
   b. sodium hyperchlorite
   c. calcium hypochlorite
   d. calcium hyperchlorite

6. In a diaphragm pump chlorinator, the amount of chlorine can be adjusted by all of the following *except* by changing the _____.
   a. pump's speed
   b. length of the pump stroke
   c. amount of time that the pump operates
   d. strength of the chlorine solution

7. In a tablet chlorinator, a _____ on the pump outlet line creates a pressure imbalance, which draws water into the tablet tank.
   a. restricting valve
   b. check valve
   c. bypass valve
   d. pressure regulator valve

8. In a pasteurization system, the electric heating chamber heats untreated water to _____°F for at least 15 seconds.
   a. 98
   b. 150
   c. 161
   d. 212

Review Questions

9. In an ultraviolet light system, a _____ allows the lamp to reach full intensity before water enters the light chamber.
   a. solenoid valve
   b. pressure switch
   c. time delay switch
   d. mercury switch

10. Most ultraviolet light units come equipped with a _____ to measure the light emitted by the lamp.
    a. photoelectric cell
    b. photovoltaic cell
    c. light receptor cell
    d. time-delay cell

3.0.0 ♦ FILTERING AND SOFTENING THE WATER SUPPLY

Unfiltered water contains particles such as sand, mud, and silt. This is called turbidity. Turbidity clogs pipes and causes excessive wear on faucets, valves, and appliances. Water can also contain chemicals such as iron, hydrogen sulfide, acids, fluoride, and vegetable tannins. Chemicals discolor water, stain fixtures and clothes, and make the water smell and taste bad. Hard water contains large amounts of mineral salts. Mineral salts coat pipes, fittings, and water heaters with scale. Unfiltered water can reduce the efficiency of a water supply system. It can also shorten the life of appliances and cause health hazards.

Water filters and water softeners separate out undesirable materials. The result is clean, safe water. Filters pass water through a porous material such as a membrane or charcoal. The porous material separates the particles and chemicals from the water. In POU systems like sand filters, the particles and chemicals are then flushed away. Other systems use cartridge filters that trap particles and chemicals in a disposable cartridge. Water softeners use chemicals to remove mineral salts from the water.

A drain and an electrical outlet should be in close proximity to the location where you install water filters and softeners. This will reduce the amount of water pipe required for the installation. The building layout will affect the placement of filters and softener piping and equipment. In basements, install the unit near the water heater (see *Figure 7*). Install shutoff valves on the inlet and outlet lines. In crawl spaces, the piping can be run under the floor (see *Figure 8*). In a slab-on-grade foundation, both the main supply and the drain will be in the floor (see *Figure 9*).

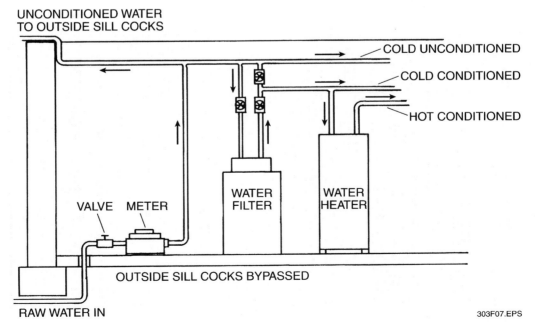

Figure 7 ♦ Water filter unit installed in a basement.

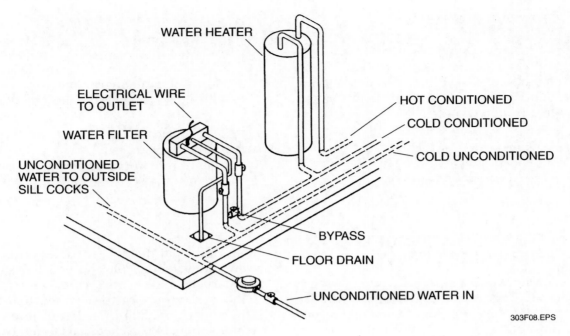

Figure 8 ◆ Water filter unit installed in a crawl space.

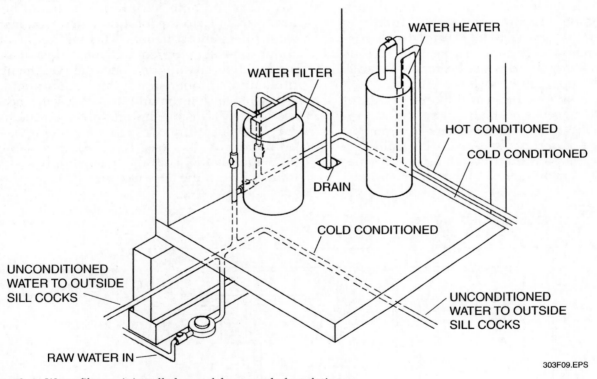

Figure 9 ◆ Water filter unit installed on a slab-on-grade foundation.

When roughing-in the piping for a water supply system, leave room for filtering and softening equipment. During the installation, try to keep contaminants out of the system. Despite your best efforts, dirt, sand, flux, pipe joint compound, metal filings, and oil will probably get inside the pipes during installation. Follow these steps to remove contaminants from a new water supply system:

Step 1   Remove the aerators and filters from all faucets in the system.

Step 2   Open all valves and faucets as wide as possible.

Step 3   Charge the system with clean water. Flush the system until the water runs clean from all faucets and valves. Disinfect the system if required.

Step 4 Reinstall all aerators and filters.

Always test the water before you install a filter or softener. The test will identify any problems with the water quality. This information will allow you to select the appropriate filter or softener. Professionals usually perform these tests. Municipal water utilities also have staff who perform water quality tests. Ask an experienced plumber to recommend a tester in your area.

You have already been introduced to the concepts of water filtering and softening. In this section, you will learn about the different ways to filter and soften water. You will also learn how to install water filters and water softeners. Refer to your local code when installing filtering and softening systems. Always take the time to diagnose the problem before you recommend a filter or softener. Both your reputation and the customer's health are at stake.

3.1.0 Municipal Water Treatment

Public water utilities treat water to ensure that it is potable. Utilities are required to follow local codes and federal water quality guidelines. Most systems can treat thousands of gallons at a time. The process begins when water is pumped to the treatment facility from a lake, river, well, or reservoir. Chemicals are added to separate out turbidity and kill harmful organisms. The water is then pumped to a settling basin, where solids sink and are filtered out. The water is then pumped through sand filters and treated with lime to remove acidity. It is then stored until it is needed.

Treated water is often stored in open reservoirs. There, it is exposed to contaminants in the atmosphere. Additional treatment is required before this water can be allowed to enter the public water supply system. The water is filtered and disinfected in a storage tank. **Diatomaceous earth** is a commonly used filter. Diatomaceous earth consists of the fossilized remains of microscopic shells created by one-celled plants, called diatoms. The fossilized remains are locked together, creating a natural filter that traps particles in the water. A chlorine and ammonia solution is added to the water to disinfect it. When the water has been filtered and disinfected, it is safe for use.

Once water leaves the public system and enters the home or office, plumbers can install additional water treatment systems as needed. Buildings supplied by a private well must have these systems. The most common systems are described below. Refer to your local code to learn which systems are permitted in your area.

3.2.0 Ion Exchange Systems

Look at the inside of an old coffeepot or teakettle. The bottom may be coated with a rough, grayish material. The material is deposits of minerals, such as calcium, magnesium, and sodium. Heat causes the minerals to settle out of the water and stick to the metal, resulting in scale. This settling is called **precipitation**. Scale can block pipes and reduce the efficiency of water heaters and other appliances. Water that contains large amounts of minerals is called hard water.

Heat causes scale by actually changing the structure of calcium, magnesium, and sodium atoms. Heat provides the atom with extra energy. The energy makes the atom shed or absorb tiny particles called electrons, giving the atom an electrical charge. The charge can be either positive or negative. Atoms with electrical charges are called **ions**. The electrical charge allows the ion to bond with a metal surface. When enough atoms bond to the surface, they become visible as scale.

Water softeners use a chemical process called **ion exchange** to remove mineral ions from hard water. Ion exchange is the process of bonding ions with other atoms. Once the ions bond with another atom, they lose their electrical charge. They then settle out of the water. One of the most common forms of ion exchange is the zeolite system (see *Figure 10*). Zeolite is a granulated chemical compound that contains iron, silica, alumina, and potash.

A zeolite system consists of two tanks, called a resin (or mineral) tank and a brine tank. Inside the resin tank are pellets of a resinous substance saturated with sodium. Hard water enters the resin tank, where the sodium bonds with the mineral ions. The resulting atoms are washed out of the tank's drain. The softened water then flows into the water supply system. The brine tank supplies sodium-rich water to replenish the resin in the resin tank.

Softeners come in a variety of designs. For fully automatic units, the owner regularly refills the brine tank with salt tablets. The unit recharges the resin tank by itself. Fully automatic units also clean themselves; this cleaning process is called **backwashing**. A timer controls the recharging and backwashing. Semiautomatic and manual units were used in the past, and are now rare.

Use portable water softeners to treat water when the water supply system is under construction and not yet tied into the municipal system. Portable softeners can be rented from local plumbing suppliers. Connect the unit to the water supply using temporary fittings. The dealer will

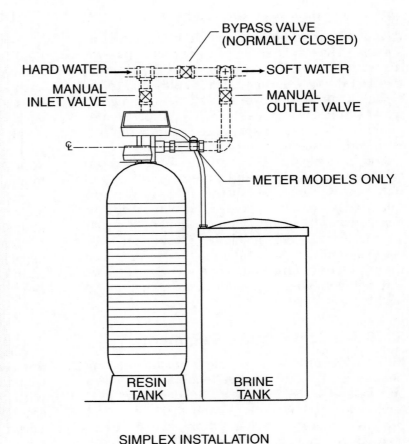

Figure 10 ♦ Zeolite water softener.

recharge the system. The customer has no equipment costs or maintenance problems. When the hard water problem clears up, unhook the unit and return it to the dealer. Ensure that the water supply system's connections are sealed properly. Ask an experienced plumber to recommend a reputable dealer in your area.

Ensure that the water softener is installed level. The softener will operate less efficiently if it is not level. Installation of a bypass valve on the supply line (see *Figure 11*) allows the softener to be removed for maintenance or replacement without disrupting the water supply. Use either a three-way combination valve or three separate valves. Refer to your local code.

Do not drain a water softener into the main waste system. Softener wastewater can contaminate a waste system. Codes require that a softener drain into a laundry tray or floor or other trapped drain. Ensure that there is an air gap of at least 1½ inches between the softener drainpipe and the receiving drain. Some codes allow softeners to drain into dry wells. Always refer to your local code before you install a softener drain.

 WARNING!
Soft water contains sodium chloride (salt). Softened water may pose a health risk to people on strict low-sodium diets. Ensure that the customer's health will not be affected by the installation of a soft water treatment system.

 WARNING!
Water heaters that are used on the same water supply line as an ion exchange system must be isolated from the ion exchange system according to your local code requirements. Otherwise chlorine can contaminate the water heater and cause sickness and even death.

After you install a water softener, disinfect it by adding chlorine to the resin tank. The chlorine will flush out any dirt and harmful organisms that may have collected in the tanks. The amount of chlorine depends on both the softener's capacity and the concentration of the chlorine. For example, you need to use 1.2 fluid ounces of domestic bleach per cubic foot of resin to disinfect a zeolite system. Use the following steps to disinfect a water softener:

Step 1 Add chlorine to the softener according to the softener's operating instructions. Allow the system to stand for 20 minutes.

Step 2 Drain the softener completely. If the drained water smells like chlorine, the system has been disinfected. If the water does not smell like chlorine, repeat Step 1.

Step 3 When the system has been disinfected, run fresh water through the softener until chlorine odor is gone.

Step 4 Charge the softener according to the operating instructions.

CAUTION

Always consult the manufacturer's recommendations for the chlorine to be used in an ion exchange system. Chlorine can affect the resin bed material.

3.3.0 Filtration Systems

There are many different types of filters. Each is designed to treat a specific water quality problem. The installation requirements for filters vary depending on type and manufacturer. Always follow the manufacturer's instructions when installing water filters. Refer to your local code for guidelines on particular types of filters.

Most filters work because of a principle called **adsorption**. When you stick a sponge in water, the sponge expands as it fills with water. This, as you know, is called absorption. Adsorption is absorption that happens only on the surface of an object. If you had an adsorbent sponge, it would not expand when placed in water. It would pick up only as much water as could stick to the outside of the sponge. While that behavior makes for a very inefficient sponge, it makes for a very efficient filter.

Activated carbon (AC) is a highly effective adsorbent. AC is very porous. That means it has many crevices in which to trap molecules. In fact, 1 pound of AC has between 60 and 150 acres of surface area. Thus, AC is ideal for adsorption. As water flows through an AC filter, molecules become trapped in the carbon. AC eliminates the organic chemicals that cause foul odor, taste, and color. It also removes chlorine and harmful chemicals, such as pesticides and solvents. AC filters do not remove bacteria and viruses, sodium, hardness, fluorides, nitrates, or heavy metals.

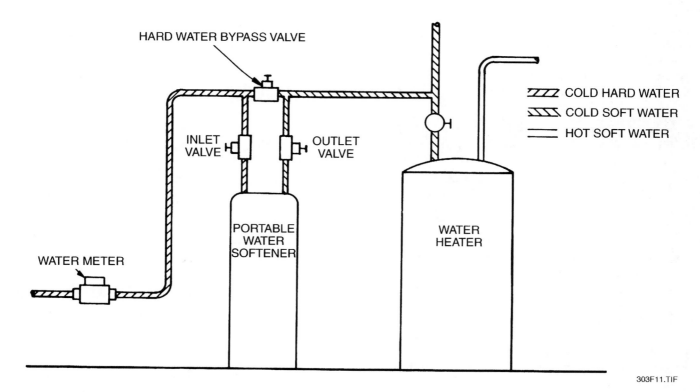

Figure 11 ♦ Installation of a bypass valve on a portable water softener.

Install AC filters at individual fixtures or on the main supply line. Low-volume units are effective for removing chemicals that cause foul odor, taste, and smell. They do not trap other materials. High-volume units have more AC and can remove harmful chemicals more effectively. Low-volume units can be installed on a faucet. Larger units should be located under a sink.

Change AC filters regularly to prevent them from clogging. Refer to the manufacturer's instructions. Many companies suggest that filters be changed after a certain amount of water has flowed through the unit. Install a sediment filter upstream of the AC filter. This will prevent the AC from clogging with silt. Some types of bacteria can grow in AC filters. They are generally harmless. Many AC filters feature inhibitors designed to prevent bacteria growth. Other common types of filters include the following:

- Mechanical filters
- Neutralizing filters
- Oxidizing filters

Mechanical filters trap large particles, such as sand, silt, and clay. They work by passing water through sand or other grainy materials. They are not effective against fine particles. Install mechanical filters along with other water treatment systems to ensure total filtration. Mechanical filters are also called microfilters.

Neutralizing filters remove acidity from water. They work by trapping the acid in material rich in calcium. Common sources of granular calcium are marble, lime, and calcium carbonate.

Oxidizing filters use zeolite treated with magnesium. The zeolite turns iron into rust, which can then be trapped in a filter. Oxidizing filters are also effective against manganese and some sulfides. Always follow the manufacturer's instructions when installing these systems. Test the water to determine the type of contamination. This will ensure that you select the correct filtration system.

3.4.0 Precipitation Systems

Extremely fine particles and certain chemicals may be able to pass through filters. These contaminants can be removed by adding precipitation as an extra step in the filtration process. Chemicals are introduced into the water to bond with the contaminants. The action of bonding is called **coagulation**. The coagulated matter is large and heavy enough to settle out of the water. The separated matter, called a **precipitate**, can then be filtered. Precipitates are also sometimes called **floc**.

Iron must be converted into much larger rust particles to be a precipitate. To turn the iron into rust, add an **oxidizing agent** to the treatment system. An oxidizing agent is a chemical that is made up largely of oxygen. It bonds with chemicals and materials that react to air. You know that exposure to air and water causes iron to rust. When the oxidizing agent bonds with iron, it turns the iron into rust. This process is called oxidation.

To filter out fine solid matter, add **alum** to the water. Alum is a chemical combination of aluminum and sulfur. The small particles coagulate onto the particles of alum. The result is larger particles that can be filtered out. For acidic water, add a soda ash solution to the water. These items are available from your local plumbing supplier. Refer to your local code for approved methods and chemicals in your area.

Many coagulants need to be in the water for a while before they can treat the water completely. To accomplish this, install a chemical feeder and retention tank. Chemical feeders inject small amounts of coagulants and chemicals into the water at regular intervals. The feeder draws the chemicals from a storage tank. A timer controls the feeder. Retention tanks allow the chemicals time to circulate through the water. A local water treatment specialist can suggest suitable tanks and feeders. Follow the manufacturer's directions when installing the equipment (see *Figure 12*).

3.5.0 Reverse Osmosis Systems

Imagine that you have a container full of water divided in half by a fine mesh. Now add a small amount of salt to the water in one half (side A) and a large amount of salt to the other (side B). You will have two separate saltwater solutions. What happens next may surprise you. See *Figure 13*.

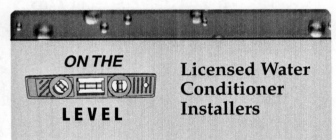

ON THE LEVEL

Licensed Water Conditioner Installers

In most states, plumbers can get a license to install water conditioners. Getting a license may not be a requirement, but it can be a big step up in your professional development. To get a license, you usually need to have experience installing and servicing conditioners. Six months' experience is a common standard. You then have to pass an exam. The exams vary from state to state. Contact your local health department to apply for an exam in your area.

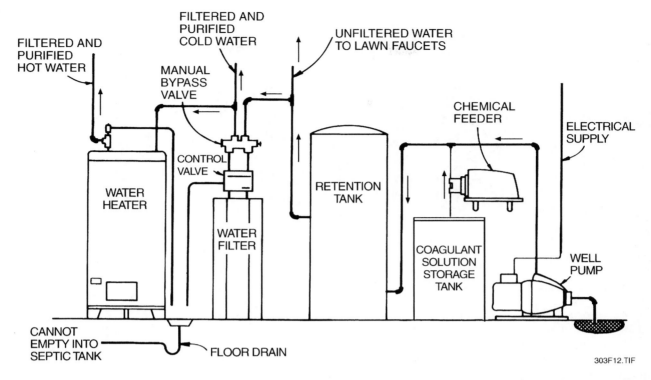

Figure 12 ◆ Installation of a chemical feeder, chemical storage tank, and retention tank in a system designed to treat well water.

Solutions tend to seek equilibrium, or balance. The water in side A will pass through the mesh to balance the stronger side B solution with more water. At the same time, the salt water in side B will try to flow through the mesh to the other side to give it more salt. The flow from side A to side B is stronger because there are many more salt particles in side B to get caught in the mesh as they try to pass through. Eventually, there will be more water in side B than side A. This unequal back-and-forth flow is called **osmosis**. The living cells in plants and animals use osmosis to obtain nutrients.

Some water treatment systems use a variation of the principle of osmosis to filter out harmful material. Using the example above, imagine that side B is subjected to an increase in pressure. The stronger solution is forced to pass through the membrane. The pressure overcomes the flow of the weaker side A solution. The membrane traps the salt as the water flows through. This flow is called reverse osmosis (see *Figure 14*).

In a reverse osmosis system, a pump forces water through a cellulose acetate membrane at a pressure of 200 to 400 pounds per square inch (psi). The holes in the membrane are small enough

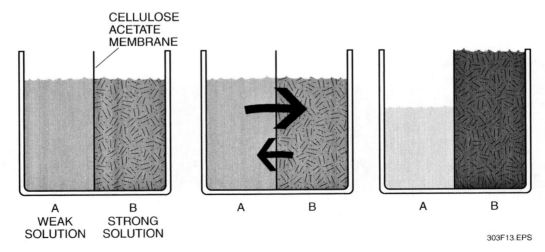

Figure 13 ◆ How osmosis works.

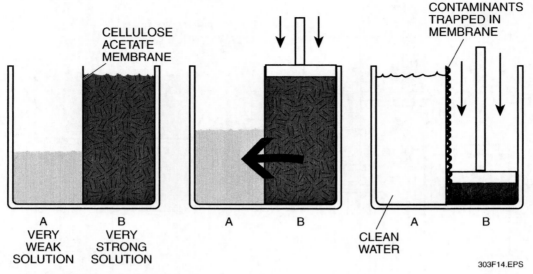

Figure 14 ◆ How reverse osmosis works.

to let only water and a few chemicals through. The other contaminants are trapped on the membrane. A directed flow of water scours the contaminants off the membrane. The contaminants are then washed away as wastewater. Reverse osmosis systems are small enough to be installed on or under a sink (see *Figure 15*).

Reverse osmosis is a very effective way to eliminate most harmful organisms and contaminants. However, it can treat only small amounts of water at a time. The process also uses a lot of water to clean the membrane. This cleaning water is directly washed out as wastewater. Depending on the installation, this can be wasteful. Codes require that all reverse osmosis systems drain into the waste system through an air gap.

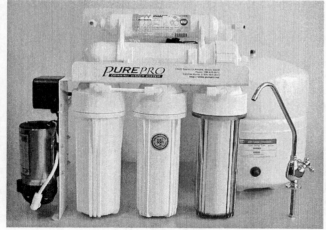

Figure 15 ◆ Reverse osmosis system.

3.6.0 Distillation Systems

Converting water to steam and back again is a very effective way to remove contaminants. As water flashes to steam, it leaves behind many impurities. This process is called **distillation**. It is one of the oldest and simplest forms of water purification. It does not require sodium, chlorine, or other chemicals. Distillation removes more than 99 percent of impurities from water, including the following:

- Bacteria
- Hardness
- Heavy metals
- Nitrates
- Organic compounds
- Sodium
- Turbidity

The body of a typical distiller is divided into a boiling chamber and a storage chamber (see *Figure 16*). Cold, untreated water is pumped through a condensing coil into the boiling chamber. There, it is heated by an electric heating element. The heat kills bacteria and viruses in the water. As the heated water turns to steam, it rises and leaves most impurities behind. The purified steam then flows past the cold condensing coil. Contact with the coil turns the steam back into a liquid. The water collects in the storage chamber, where it is fed into a storage tank for use. The contaminated water that remains in the boiling chamber is flushed into a drain.

Distillers are also called stills. They can usually process only small amounts of water at a time. Typical home units process 3 to 11 gallons per day.

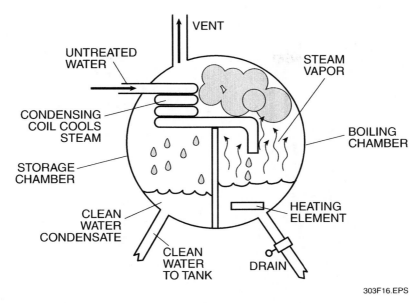

Figure 16 ◆ Diagram of a typical distillation unit.

They use more energy than other treatment systems. Also, some chemicals, such as benzene and toluene, have a lower boiling point than water. They are called volatile organic compounds (VOCs). VOCs can be trapped in steam. They can remain trapped in the water after it cools. Most distillers have filters or vents to allow VOCs to escape. Refer to your local code for guidelines on distillers.

Distillers may be installed at individual fixtures. Ensure that the distiller is properly vented and that all filters are clean. Bacteria can collect around the cooling coils when the distiller is not in use. Ensure that the coils are clean before each use. Scale may form around the heating element. Clean the heating element regularly. Vinegar is effective for cleaning heating elements. Follow the manufacturer's cleaning instructions. Some harsh cleaners can damage the distiller's casing. With proper care, a home distiller can last up to 15 years.

4.0.0 ◆ TROUBLESHOOTING WATER SUPPLY PROBLEMS

The following are the most commonly encountered problems in water supply systems:

- Hardness
- Discoloration
- Acidity
- Bad odors and flavors
- Turbidity

In this section, you will learn how to identify and correct each of these problems. You will learn how to identify the causes and symptoms of the different problems and how to test the water to confirm your diagnosis. The measurements and treatments discussed may vary depending on the local code.

Always refer to your local code before you attempt to correct a water treatment problem. Consult with expert plumbers about common water supply problems in your area, especially while you are still learning on the job. Later, as you become more experienced, you can return the favor to apprentices.

4.1.0 Hardness

As you have learned already, hardness is caused by calcium and magnesium salts in the water. The salts may be in the form of bicarbonates, sulfates, or chlorides. Iron salts can also contribute to hardness. When soap is added to hard water, a sticky curd will form on the surface of the water. Suds do not form readily in hard water, and your skin feels rough after you wash in it. Hard water will cause glassware to appear streaked and murky. It also forms hard, scaly deposits on the inside of metal pipes.

Water hardness is measured in **grains per gallon** of hardness. One grain equals 17.1 ppm of hardness. To determine the degree of water hardness, draw a 2-ounce sample of the water. Add one drop of a soap solution to the water and shake it to form lasting suds. If no suds form, continue to add one drop at a time and shake until lasting suds form. The number of drops used is approximately equal to the hardness in grains per gallon. Consult *Table 1* to interpret the test results. Water with more than 3 grains of hardness requires the installation of a zeolite softener or a reverse osmosis unit.

Table 1 Measure of Water Hardness Using Soap Solution Test

Quantity of Soap Solution	Hardness
0–3 drops (0–3 grains per gallon) (approximately 50 ppm)	Soft
4–6 drops (4–6 grains per gallon) (50–100 ppm)	Moderately hard
7–12 drops (7–12 grains per gallon) (100–200 ppm)	Hard
13–18 drops (13–18 grains per gallon) (approximately 200–300 ppm)	Very hard
19 or more drops (19 or more grains per gallon) (over 300 ppm)	Extremely hard

4.2.0 Discoloration

Iron and manganese particles dissolved in the water can cause discoloration. Sometimes the presence of iron- and manganese-eating bacteria has the same effect. Dissolved iron turns water red. Iron-contaminated water will stain clothes and porcelain plumbing fixtures, even at concentrations as low as 0.3 ppm, and will corrode steel pipe. Water with dissolved iron has a metallic taste. Freshly drawn water may appear clear at first, but after exposure to the air, iron particles will gradually settle out.

Test kits are available to test for iron contamination. You can purchase a test kit from a local plumbing supplier. The kit contains an acid solution, a color solution, and a color chart. Use the following steps to test the water for the presence of iron:

Step 1 Draw a water sample according to the instructions in the test kit. Add the acid solution to the sample according to the test instructions. The acid will dissolve any iron that has settled out of the water.

Step 2 Add the color solution to the water sample according to the test instructions. The water sample will turn a shade of pink.

Step 3 Match the color of the water sample against the standard color chart in the test kit. The chart will indicate the amount of iron in the water based on the color. Charts are usually rated in ppm.

Step 4 Install a water conditioner if the test indicates the presence of more than 0.2 ppm of iron in the water (see *Table 2*).

Table 2 lists the various treatments for different levels of iron contamination. Note that there may be more than one treatment option for a given level of contamination. The selection will depend on product availability, installation and maintenance costs, and the plumber's familiarity with the systems.

Iron-eating bacteria are often associated with the presence of acid or other corrosive conditions. These bacteria create a red, slippery jelly on the inner surfaces of toilet tanks. To eliminate iron bacteria, heavily chlorinate the water source, pump, and piping system. This will kill the bacteria that are already in the system. Install a chlorinator and filter to prevent subsequent buildups.

Brownish-black water indicates the presence of both manganese and iron. Manganese and iron can stain fixtures and fabrics. They can also give a bitter taste to coffee and tea. Manganese bacteria also cause a slippery buildup on the inside of toilet tanks, but of a darker color. Iron test kits also detect manganese. Treat manganese contamination in the same way as iron contamination.

4.3.0 Acidity

Decaying vegetable matter emits carbon dioxide gas. Carbon dioxide can also be found in the atmosphere. When water absorbs and reacts with carbon dioxide, a weak acid solution is created. This acid slowly eats away at metal pipes and the metal components in fixtures, tanks, and pumps. In rare instances, some water may contain sulfuric, nitric, or hydrochloric acids. These can cause damage even more quickly. If copper or brass pipes are being eaten by acid, the water may leave

Table 2 Treatment Recommendations for Iron-Contaminated Water

Iron Concentration (ppm)	Treatment
0–0.2	No treatment necessary
1–2	Polyphosphate feeder
0.2–10	Zeolite softener
1–10	Oxidizing (Manganese zeolite) filter, usually manganese-treated green sand
3 or more	Chlorinator and filter, usually a sand or carbon filter

green stains on fixtures and faucets. Acids also reduce the effectiveness of iron treatments.

Acid test kits are available from your local plumbing supplier. They contain a chemical indicator and a color chart similar to those found in iron test kits. Draw a water sample and add the chemical indicator to it. Compare the color of the water with the kit's color chart. The colors are keyed to the level of acidity in the water, usually by pH level. Decreasing numbers indicate increasing acidity. If the pH is 7 or below, the water supply should be treated for acidity.

Acidity can be neutralized by using a soda ash solution. Feed the solution into the water supply system using one of the following methods:

- Directly into a well
- Through a well pump suction line
- Along with a chlorine solution
- Through a separate feeder unit

A neutralizing tank with limestone or marble chips will also remove acidity from the water. Note that this process will increase the water's hardness. Always install a water softener on the line when installing a neutralizing tank. The softener must be downstream of the tank to be effective.

4.4.0 Foul Odors and Flavors

The following contaminants cause rotten egg odor and flavor in water:

- Hydrogen sulfide gas
- Sulfate-reducing bacteria
- Sulfur bacteria

These contaminants eat metal in pumps, piping, and fixtures. If the water contains sulfur and iron, fine black particles may collect in fixtures. This condition is commonly called black water, and it can turn silverware black. Do not cook with water that has been contaminated by sulfur.

Consult with a water-conditioning equipment dealer on the level of sulfur contamination. The dealer will be able to test the water on the spot. If the water contains more than 1 ppm of sulfur, it should be treated. The water can be treated using chlorinators and filters. Manganese-treated green sand filters can also be used for contamination levels of 5 ppm or less.

The following factors can also cause off flavors:

- Extremely high mineral content
- Presence of organic matter
- Excess chlorine
- Passage of water through areas with oily or salty waste

These contaminants cause water to taste bitter, brackish, oily, or salty. They can also give a chlorine odor or taste to the water. You can purchase tests for each of these contaminants from your local plumbing supplier. Install a water softener to eliminate metallic flavors. Salty flavors can be treated with reverse osmosis units. Activated carbon filters are effective against most other tastes and odors, including chlorine.

4.5.0 Turbidity

Salt, sediment, small organisms, and organic matter give water a dirty or muddy appearance, called turbidity. Turbidity is generally measured in **turbidity units**. A turbidity unit is a measure of the amount of light that can pass through a water sample. Turbidity can range from less than 1 ppm up to 4,000 ppm. Turbidity tests require special laboratory equipment. Contact your local health department or a filter manufacturer to arrange for a test. Water should be treated if its turbidity is greater than 10 ppm. Use a sand or diatomaceous earth filter. Other options include surface treatment of the water source with powdered gypsum or copper sulfate and installation of a settling tank filled with alum.

Review Questions

Sections 3.0.0–4.0.0

1. Water that has been stored outside at a municipal water treatment facility is often filtered through _____ before it is used.
    a. diatomaceous earth
    b. a mixture of sand and clay
    c. activated charcoal
    d. cellulose acetate membranes

2. An ion exchange system _____ the electric charge of ions in the water.
    a. increases
    b. decreases
    c. reverses
    d. eliminates

3. To allow coagulants enough time to circulate in untreated water, install a(n) _____ and a retention tank.
    a. solenoid valve
    b. activated carbon filter
    c. chemical feeder
    d. restricting valve

4. Osmosis is the flow of a _____ solution to a _____ solution through a membrane.
    a. lighter; heavier
    b. stronger; weaker
    c. heavier; lighter
    d. weaker; stronger

5. Regular maintenance on a distiller includes all of the following *except*:
    a. clean heating element regularly
    b. flush the system with bleach
    c. ensure that it is properly vented
    d. clean the coils before each use

6. One grain of hardness equals _____ ppm of hardness.
    a. 7.25
    b. 12.1
    c. 16.5
    d. 17.1

7. Treat a water supply system for acidity if the pH value is _____ or below.
    a. 7
    b. 9
    c. 11
    d. 13

8. The presence of fine black particles in fixtures indicates contamination by _____ and _____.
    a. manganese; sulfur
    b. iron; chlorine
    c. sulfur; iron
    d. sulfides; manganese

9. A turbidity unit is a measure of the amount of _____ that can pass through a water sample.
    a. sediment
    b. electricity
    c. light
    d. heat

10. Water should be treated if its turbidity is _____ ppm or greater.
    a. 5
    b. 10
    c. 15
    d. 20

Summary

Plumbers install water conditioners to provide safe, clean water. Conditioners remove harmful organisms and materials from the water. Water can be treated by disinfection, filtration, and softening. Disinfection kills bacteria and viruses. Filtration removes particles and chemicals. Softening removes the chemicals that can cause scale. These three methods can be used separately or together.

Disinfection kills those organisms that can cause sickness and death. Plumbers disinfect new water supply systems when they are completed. Common disinfection methods include chlorination, pasteurization, and ultraviolet light. Chlorine, found in bleach, kills harmful organisms in water. It also eliminates odors and foul taste. Chlorine can be used in any size water system. Pasteurization heats water enough to kill any living creatures that are in the water. This method does not remove other harmful products from the water and does not treat large amounts of water at a time. Ultraviolet light also kills organisms using heat. A powerful lamp heats the water. This method does not give water a chemical taste. However, minerals in the water can coat the lamp and reduce its efficiency.

Filters remove particles and chemicals from water. Most filters work on the principle of adsorption. Adsorption is a type of absorption that happens only on the surface of an object. Activated carbon (AC) is an effective adsorbent and is a common form of water filter. When water flows through AC, particles are trapped in the pores of the carbon. Mechanical filters trap sand, silt, and clay by passing water through sandy material. This method works best against coarse particles. Neutralizing filters trap acid wastes in marble, lime, or calcium carbonate. Oxidizing filters turn iron into rust.

This module addressed other common forms of filtration, including precipitation, reverse osmosis, and distillation. Precipitation is an extra step that involves adding a chemical to the water. The chemical bonds with fine particles and chemicals in the water. The resulting large particle is called a precipitate. Precipitates are large enough to be caught by filters. Reverse osmosis forces water under pressure through a cellulose acetate membrane. The membrane traps everything except water and a few organic chemicals. This method is a very effective way to treat water. Distillation treats water by boiling it. When water turns to steam, it leaves most contaminants behind. The clean steam is then cooled and stored as water.

Minerals that cause scale can be removed from water using an ion exchange system. This system neutralizes the electrical charge of the particles that cause scale. The neutralized particles can then be filtered out. Zeolite systems are the most popular form of ion exchange. Zeolite is a chemical compound made up of iron, silica, alumina, and potash.

This module reviewed commonly encountered problems in water supply systems, including hardness, discoloration, acidity, bad odors and flavors, and turbidity. Understanding how to identify and treat these problems is an essential part of troubleshooting. Measurements and treatments vary depending on location, so always refer to your local code.

Hardness is caused by calcium and magnesium salts in the water. Install a zeolite softener or reverse osmosis unit to correct hardness. Iron and magnesium can cause discoloration, making the water appear red or taste metallic. A test kit can confirm the diagnosis and help determine the appropriate treatment. Acidity eats away at metal pipes and components, leaving green stains on fixtures and faucets. Neutralize the water with a soda ash solution to resolve acidity. Water that smells or tastes of rotten egg can indicate the presence of sulfur. Such water can be treated using chlorinators and filters. Salt, sediment, and organic matter create a muddy appearance, which can be an indication of high turbidity. Special laboratory equipment is required to conduct turbidity tests.

Customers expect to have clean, safe water for drinking, washing, and cleaning. You can ensure that a customer's water is safe by testing it. The test will show the types of contaminants that are in the water. You can then install water conditioners that will remove the problem. Learn to identify the organisms and materials that can pollute a water supply system. With experience, you will be able to provide systems that suit customers' needs.

Trade Terms Introduced in This Module

Adsorption: A type of absorption that happens only on the surface of an object.

Alum: A chemical used for coagulation. It consists of aluminum and sulfur.

Backwashing: The cleaning process of a fully automatic water softener.

Calcium hypochlorite: The solid form of chlorine. It comes in powder and tablet form.

Chlorination: The use of chlorine to disinfect water.

Chlorinator: A device that distributes chlorine in a water supply system.

Coagulation: The bonding that occurs between contaminants in the water and chemicals introduced to eliminate them.

Contamination: An impairment of the quality of the potable water which creates an actual hazard to the public health through poisoning or through the spread of disease by sewage, industrial fluids, or waste. Also defined as High Hazard. (As defined in the 2003 *Uniform Plumbing Code*.)

Diaphragm pump chlorinator: A type of chlorinator that uses a diaphragm pump to circulate chlorine.

Diatomaceous earth: Soil consisting of the microscopic skeletons of plants that are all locked together.

Disinfection: The destruction of harmful organisms in water.

Distillation: The process of removing impurities from water by boiling the water into steam.

Floc: Another term for precipitate.

Grains per gallon: A measure of the hardness of water. One grain equals 17.1 ppm of hardness.

Injector chlorinator: A type of chlorinator that uses an injector to circulate chlorine.

Ion: An atom with a positive or negative electrical charge.

Ion exchange: A technique used in water softeners to remove hardness by neutralizing the electrical charge of the mineral atoms.

Osmosis: The unequal back-and-forth flow of two different solutions as they seek equilibrium.

Oxidizing agent: A chemical made up largely of oxygen used for oxidation.

Pasteurization: Heating water to kill harmful organisms in it.

Point-of-entry unit: A water conditioner that is located near the entry point of a water supply.

Point-of-use unit: A water conditioner that is located at an individual fixture.

Pollution: An impairment of the quality of the potable water to a degree which does not create a hazard to the public health but which does adversely and unreasonably affect the aesthetic qualities of the potable water for domestic use. Also defined as Low Hazard. (As defined in the 2003 *Uniform Plumbing Code*.)

Precipitate: A particle created by coagulation that settles out of the water.

Precipitation: The process of removing contaminants from water by coagulation.

Sodium hypochlorite: The liquid form of chlorine, commonly found in laundry bleach.

Solenoid valve: An electronically operated plunger used to divert the flow of water in a water conditioner.

Stroke: A single operating cycle of a diaphragm pump.

Tablet chlorinator: A type of chlorinator in which water is pumped through a container of calcium hypochlorite tablets.

Turbidity unit: A measurement of the percentage of light that can pass through a sample of water.

Ultraviolet light: Type of disinfection in which a special lamp heats water as it flows through a chamber.

Water conditioner: Device used to remove harmful organisms and materials from water.

Resources & Acknowledgments

Additional Resources

This module is intended to be a thorough resource for task training. The following reference works are suggested for further study. These are optional materials for continued education rather than for task training.

Basic Principles of Water Treatment. 1996. Cliff Morelli. Littleton, CO: Tall Oaks Publishing.

Practical Principles of Ion Exchange Water Treatment. 1985. Dean L. Owens. Littleton, CO: Tall Oaks Publishing.

Water Treatment Specification Manual. 1993. Frank Rosa. New York, NY: McGraw-Hill.

References

Dictionary of Architecture and Construction, Third Edition. 2000. Cyril M. Harris, ed. New York: McGraw-Hill.

2003 International Plumbing Code. International Code Council. Falls Church, VA: International Code Council.

2003 Uniform Plumbing Code. International Association of Plumbing and Mechanical Officials. Ontario, CA. International Association of Plumbing and Mechanical Officials. (For definitions of the trade terms "Contamination" [section 205.0] and "Pollution" [section 218.0].)

Michigan State University Extension Service Website, http://www.msue.msu.edu, Bulletin WQ22, "Distillation for Home Water Treatment," www.msue.msu.edu/portal/default.cfm?pageset_id=25744&page_id=25794&msue_portal_id=25643 (search by "Publication," enter inventory number "WQ22"), reviewed September 2004.

North Dakota State University Extension Service Website, http://www.ext.nodak.edu, Publication AE-1029, "Treatment Systems for Household Water Supplies: Activated Carbon Filtration," February 1992, http://www.ext.nodak.edu/extpubs/h2oqual/watsys/ae1029w.htm, reviewed September 2004.

North Dakota State University Extension Service Website, http://www.ext.nodak.edu, Publication AE-1032, "Treatment Systems for Household Water Supplies: Distillation," April 1992, http://www.ext.nodak.edu/extpubs/h2oqual/watsys/ae1032w.htm, reviewed September 2004.

Pipefitters Handbook. 1967. Forrest R. Lindsey. New York: Industrial Press Inc.

Figure Credits

American Association for Vocational Instructional Materials, 303F02–303F06

Culligan International, 303F10

Pure-Pro USA Corporation, 303F15

CONTREN® LEARNING SERIES — USER UPDATE

The NCCER makes every effort to keep these textbooks up-to-date and free of technical errors. We appreciate your help in this process. If you have an idea for improving this textbook, or if you find an error, a typographical mistake, or an inaccuracy in NCCER's Contren® textbooks, please write us, using this form or a photocopy. Be sure to include the exact module number, page number, a detailed description, and the correction, if applicable. Your input will be brought to the attention of the Technical Review Committee. Thank you for your assistance.

Instructors – If you found that additional materials were necessary in order to teach this module effectively, please let us know so that we may include them in the Equipment/Materials list in the Annotated Instructor's Guide.

Write: Product Development and Revision
National Center for Construction Education and Research
P.O. Box 141104, Gainesville, FL 32614-1104

Fax: 352-334-0932

E-mail: curriculum@nccer.org

Craft _____ Module Name _____

Copyright Date _____ Module Number _____ Page Number(s) _____

Description
_____
_____
_____
_____

(Optional) Correction
_____
_____
_____

(Optional) Your Name and Address
_____
_____
_____

Plumbing Level Three

02304-06

Backflow Preventers

02304-06
Backflow Preventers

Topics to be presented in this module include:

1.0.0	Introduction	4.2
2.0.0	Backflow and Cross-connections	4.2
3.0.0	Types of Backflow Preventers	4.5
4.0.0	Specialty Backflow Preventers	4.13
5.0.0	Backflow Preventer Testing	4.14

Overview

Codes, health departments, and water purveyors all require plumbing systems to provide protection against reverse flow, or backflow. Backflow preventers help keep drinking water safe. Many local codes now require specific backflow preventer types for most fixtures in a new structure. Plumbers are responsible for installing backflow preventers in plumbing systems. They should know how backflow can happen, what causes back pressure and back siphonage, and the hazardous possibilities of cross-connections.

Backflow protection can be in the form of a gap or a barrier between two sources of flow. Each of the six basic types of backflow preventers is designed to work under different conditions. They range from a simple air gap, required in all fixture faucets, to a reduced-pressure zone principle backflow preventer assembly, the most sophisticated type. Backflow preventers are classified according to whether they can be used in low-hazard or high-hazard applications. For applications like hospitals and laboratories, specialty backflow preventers are used to prevent against back siphonage of dangerous chemicals and biological hazards.

Before installing a backflow preventer, plumbers must always ensure that the type selected is appropriate for the type of installation. They should consider the risk of cross-connection, the type of backflow being protected against, and the health risk posed by contaminants. Plumbers should be familiar with their local applicable code, which includes tables or charts that specify the approved applications for backflow preventers. Only certified technicians, who may or may not be plumbers, are allowed to service and test backflow preventers.

⌐ Focus Statement

The goal of the plumber is to protect the health, safety, and comfort of the nation job by job.

⌐ Code Note

Codes vary among jurisdictions. Because of the variations in code, consult the applicable code whenever regulations are in question. Referring to an incorrect set of codes can cause as much trouble as failing to reference codes altogether. Obtain, review, and familiarize yourself with your local adopted code.

Portions of this publication reproduce tables and figures from the *2003 International Plumbing Code* and the *2000 IPC Commentary*, International Code Council, Inc., Falls Church, Virginia. Reproduced with permission. All rights reserved.

Objectives

When you have completed this module, you will be able to do the following:

1. Explain the principle of backflow due to back siphonage or back pressure.
2. Explain the hazards of backflow and demonstrate the importance of backflow preventers.
3. Identify and explain the applications of the six basic backflow prevention devices.
4. Install common types of backflow preventers.

Trade Terms

Atmospheric vacuum breaker
Back siphonage
Backflow
Cross-connection
Double-check valve assembly (DCV)
Dual-check valve backflow preventer assembly (DC)
High hazard
Hose connection vacuum breaker
In-line vacuum breaker
Intermediate atmospheric vent vacuum breaker
Low hazard
Manufactured air gap
Pressure-type vacuum breaker
Reduced-pressure zone principle backflow preventer assembly
Test cock

Required Trainee Materials

1. Appropriate personal protective equipment
2. Sharpened pencils and paper
3. Copy of local applicable code
4. Calculator

Prerequisites

Before you begin this module, it is recommended that you successfully complete *Core Curriculum; Plumbing Level One; Plumbing Level Two; Plumbing Level Three*, Modules 02301-06 through 02303-06.

This course map shows all of the modules in the third level of the *Plumbing* curriculum. The suggested training order begins at the bottom and proceeds up. Skill levels increase as you advance on the course map. The local Training Program Sponsor may adjust the training order.

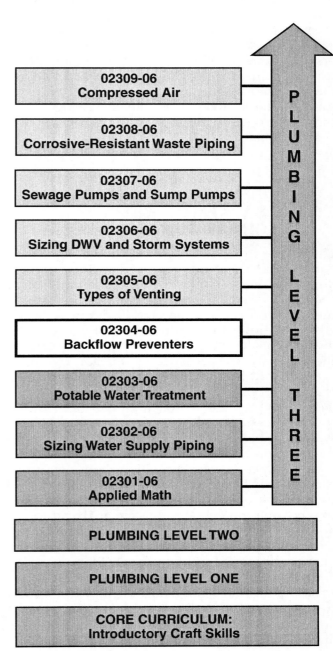

1.0.0 ◆ INTRODUCTION

Plumbing systems deliver fresh water and take away wastewater. Normally, these two water sources are separated. However, sometimes problems or malfunctions can force water to flow backward through a system. When this happens, wastewater and other liquids can be siphoned into the fresh water supply. This can cause contamination, sickness, and even death.

Codes, health departments, and water purveyors all require plumbing systems to provide protection against reverse flow. Protection can be in the form of a gap or a barrier between the two sources. Check valves can be effective barriers and can be arranged to provide increasing protection against contamination. In this module, you will learn how to install devices that protect potable water from pollution and contamination.

2.0.0 ◆ BACKFLOW AND CROSS-CONNECTIONS

The reverse flow of nonpotable liquids into the potable water supply is called **backflow**. Backflow can contaminate a fixture, a building, or a community. There are two types of backflow: back pressure and **back siphonage**. Back pressure is any elevation of pressure in the downstream direction of a potable water system that increases the downstream pressure to the point where reversal of flow direction occurs. Back siphonage is a form of backflow that, due to a reduction in the system pressure upstream, causes a subatmospheric pressure to exist at any site in the water system and results in siphoning of water from downstream.

Backflow cannot happen unless there is a direct connection between the potable water supply and another source. This condition is called a **cross-connection.** Cross-connections are not hazards all by themselves (see *Figure 1*). In fact, sometimes they are required. Cross-connections are sometimes hard for the public to spot. Many people might not know that a hose left in a basin of wastewater could cause a major health hazard by creating a backflow. Consider the cross-connection in *Figure 1*. If someone opens the faucet while the washing machine is running, the wastewater in the sink would be forced by the higher pressure back into the potable water supply, contaminating the building supply and possibly even the municipal water supply system.

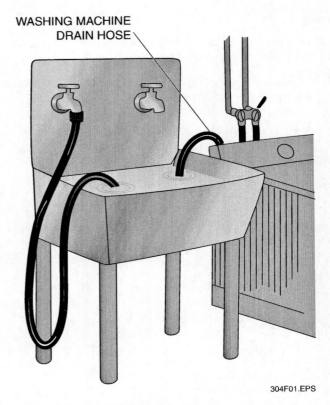

Figure 1 ◆ Simple cross-connection.

When something creates a backflow, a cross-connection becomes dangerous. Several things can cause backflow, which, in turn, can force wastes through a cross-connection:

- Cuts or breaks in the water main (see *Figure 2*)
- Failure of a pump
- Injection of air into the system
- Accidental connection to a high-pressure source

Plumbers use backflow preventers to protect against cross-connections. They provide a gap or a barrier to keep backflow from entering the water supply. Many local codes now require certain types of backflow preventers for most fixtures in a new structure, while an air gap is required for others. Tank trucks filled with sewage, pesticides, or other dangerous substances must also use backflow preventers. Plumbers are responsible for installing backflow preventers in plumbing systems. Only certified people, who may or may not be plumbers, are allowed to service them.

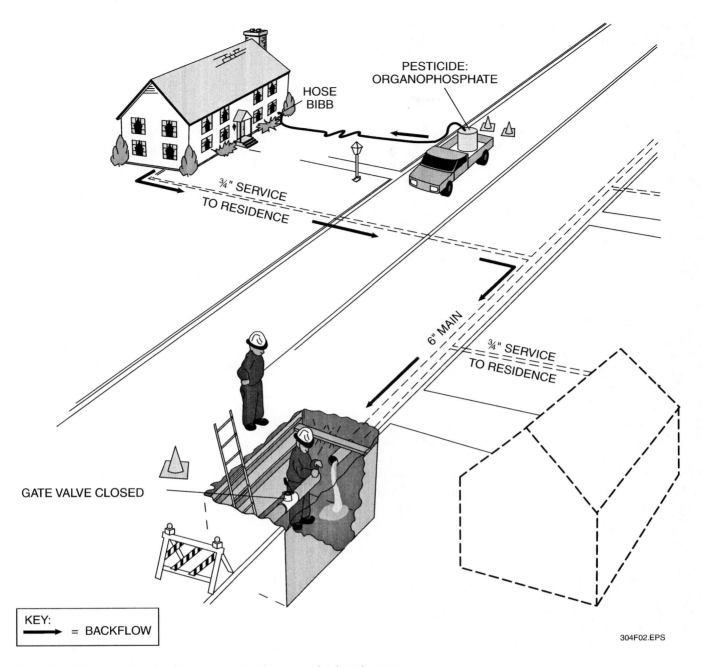

Figure 2 ◆ How a cut in a water main creates dangerous back siphonage.

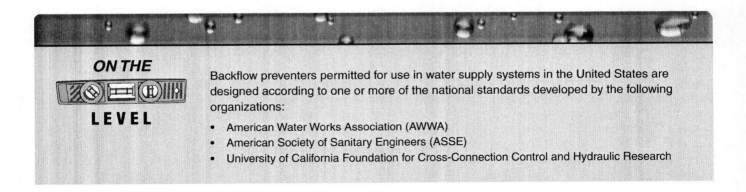

ON THE LEVEL

Backflow preventers permitted for use in water supply systems in the United States are designed according to one or more of the national standards developed by the following organizations:

- American Water Works Association (AWWA)
- American Society of Sanitary Engineers (ASSE)
- University of California Foundation for Cross-Connection Control and Hydraulic Research

DID YOU KNOW?
Backflow Blotter

The United States Environmental Protection Agency (EPA) tracks reports of backflow contamination. Here is a small sampling of the reports it has collected over the years. They serve as a warning about what can happen when people don't take proper precautions. Don't let one of your projects end up in its next report!

Seattle, Washington, 1979: A pump used in the scrubbing cycle at a car wash broke down. To keep the scrubber working, the staff ran a hose from the scrubber cycle to the rinse cycle. The rinse cycle was connected to the potable water system. Two days later, a plumber fixed the pump. However, no one remembered to remove the hose that connected the scrubbers to the potable water system. The new high-pressure scrubber pump created back pressure. This forced wastewater into the fresh water system. An eight-block area was contaminated, and two people got sick. The city ordered the car wash owner to install a reduced-pressure zone principle backflow preventer, which he did the following day.

Unidentified town, Connecticut, 1982: Workers repairing a tank at a propane storage facility purged the tank with water. They used a hose connected to a fire hydrant. However, pressure in the tank was about 20 psi higher than that of the water system. About 2,000 cubic feet of propane gas—enough to fill 1 mile of an 8-inch-diameter water main—backflowed into the water system. The city was alerted by reports of gas hissing out of fixtures. Hundreds of homes and businesses had to be evacuated. Two homes caught fire, and one of them was completely gutted. At another house, a washing machine exploded.

Woodsboro, Massachusetts, 1983: Someone working at an agricultural farm left a gate valve to a holding tank open, creating a cross-connection. The tank was filled with a powerful herbicide. About the same time, a pump failed in the town's water system, reducing the system's pressure. The herbicide from the tank siphoned into the potable water supply. People reported that yellow water was coming from their fixtures. Firefighters warned people to avoid drinking, cooking, or bathing with the town's water. The town had to flush the entire water system. Tanker trucks had to be brought in to supply fresh water, and sampling and testing had to be carried out for several weeks.

Lacey's Chapel, Alabama, 1986: The driver of a tanker truck flushed his truck's tank with water. The tank had been filled with the chemical sodium hydroxide. The driver filled the tank from a connection at the bottom of the truck. There was no backflow preventer at the site where the driver filled his truck. At the same time, the town's water main broke. Because of the sudden drop in pressure, the contents of the tanker truck siphoned into the system. The city water service did not have a cross-connection program. As a result, the wastewater entered the potable water supply. People reported burned mouths and throats after drinking the water. One man suffered red blisters all over his body after showering in the water. He and several others were treated at a nearby emergency room. Sixty homes were contaminated. The mains had to be flushed, and health officials had to inspect all homes in the affected area.

Groveton, Pennsylvania, 1981: A building was being treated for termites with two toxic pesticides: chlordane and heptachlor. The pesticide contractor diluted the mixture by running water from a hose bibb to his chemical tanks (refer to *Figure 2*). The end of the hose was submerged in the chemical solution, which set up the link for the cross-connection. At the same time, a plumber cut into a 6-inch main line to install a gate valve. As the line was cut and the water drained, a siphoning effect was created. This drew the toxic chemicals into the water main. When the water main was recharged, these chemicals were dispersed throughout the potable water system beyond the newly installed gate valve. As a result, the potable water system in 75 housing units was contaminated. Continued flushing of the water lines failed to remove the chemicals from the water supply. All of the water supply piping had to be replaced, which cost $300,000. A simple hose connection vacuum breaker could have prevented this costly mistake.

Review Questions

Sections 1.0.0–2.0.0

1. Backflow is the _____ of nonpotable liquids into the potable water supply.
   a. insertion
   b. reverse flow
   c. pumping
   d. draw off

2. Back pressure is caused by _____.
   a. an elevation of pressure in the downstream piping system
   b. equal upstream pressure in the piping system
   c. water main breaks in the downstream piping system
   d. low flow from upstream in the piping system

3. Back siphonage is caused by _____.
   a. uneven downstream pressure in the piping system
   b. leaking pipes in the piping system
   c. equal upstream pressure in the piping system
   d. a reduction in the system pressure upstream

4. One common cause of backflow is _____.
   a. failure of check valves
   b. malfunctioning pumps
   c. clogged drains
   d. improperly located vents

5. Backflow preventers consist of a gap or _____ that prevents backflow in case of a cross-connection.
   a. relief valve
   b. barrier
   c. trap
   d. intercept

3.0.0 ◆ TYPES OF BACKFLOW PREVENTERS

Backflow preventers help keep drinking water safe. They are an important part of a plumbing system because they help maintain public health. There are six types of basic backflow preventers, each designed to work under different conditions. Some protect against both back pressure and back siphonage, while others can handle only one type of backflow. Before installing a backflow preventer, always ensure that it is appropriate for the type of installation. Consider several factors:

- Risk of cross-connection
- Type of backflow
- Health risk posed by contaminants

Preventers must be installed correctly. Select the proper preventer for the application, and review the manufacturer's instructions before installing a preventer. Never attempt to install a backflow preventer in a system for which it is not designed because it may fail, resulting in backflow. Refer to your local code for specific guidance.

Backflow preventers are classified according to whether they are used in **low-hazard** or **high-hazard** applications. Both "low-hazard" and "high-hazard" refer to the possibility of a threat to the quality of the potable water supply. The difference is that a low-hazard application does not threaten to create a public health hazard although the pollutants would adversely and unreasonably affect the appearance, taste or other qualities of the water. In a high-hazard application, the potential for water quality impairment would create an actual hazard to public health through poisoning or through the spread of disease.

Your local applicable code will include tables or charts that specify the approved applications for backflow preventers. Ensure that the backflow preventer you install is suitable for the hazard as specified in your local applicable code. The safety and health of the public depends on it.

ON THE LEVEL

Low-Hazard and High-Hazard Applications

Plumbing codes classify backflow preventers according to whether they can be used in low- or high-hazard applications. However, individual codes often vary on whether certain types of backflow preventers are suitable for different hazards. As one example, the *2003 International Plumbing Code* specifies:

Device	Degree of Hazard[a]	Application[b]
Air gap	High or low hazard	Back siphonage or back pressure
Air gap fittings for use with plumbing fixtures, appliances and appurtenances	High or low hazard	Back siphonage only
Antisiphon-type fill valves for gravity water closet flush tanks	High hazard	Back siphonage only
Barometric loop	High or low hazard	Back siphonage only
Reduced-pressure principle backflow preventer and reduced-pressure principle fire protection backflow preventer	High or low hazard	Back pressure or back siphonage Sizes ⅜"–16"
Reduced-pressure detector fire protection backflow prevention assemblies	High or low hazard	Back siphonage or back pressure (fire sprinkler systems)
Double-check backflow prevention assemby double-check fire protection backflow prevention assembly	Low hazard	Back pressure or back siphonage Sizes ⅜"–16"
Double-check detector fire protection backflow preventer assemblies	Low hazard	Back pressure or back siphonage (fire sprinkler systems) Sizes 1½"–16"
Dual-check-valve-type backflow preventer	Low hazard	Back pressure or back siphonage Sizes ¼"–1"
Backflow preventer with intermediate atmospheric vents	Low hazard	Back pressure or back siphonage Sizes ¼"–¾"
Backflow preventer for carbonated beverage machines	Low hazard	Back pressure or back siphonage Sizes ¼"–⅜"
Pipe-applied atmospheric-type vacuum breaker	High or low hazard	Back siphonage only Sizes ¼"–4"
Pressure vacuum breaker assembly	High or low hazard	Back siphonage only Sizes ½"–2"
Hose-connection vacuum breaker	High or low hazard	Low head back pressure or back siphonage Sizes ½", ¾", 1"
Vacuum breaker wall hydrants, frost-resistant, automatic draining type	High or low hazard	Low head back pressure or back siphonage Sizes ¾", 1"
Laboratory faucet backflow preventer	High or low hazard	Low head back pressure and back siphonage
Hose connection backflow preventer	High or low hazard	Low head back pressure, rated working pressure back pressure, or back siphonage Sizes ½"–1"
Spill-proof vacuum breaker	High or low hazard	Back siphonage only Sizes ¼"–2"

[a] See your local code for definitions of low and high hazard.
[b] See your local code for definitions of back pressure; back pressure, low head; and back siphonage.

Remember, always consult your local applicable code to determine whether a backflow preventer is suitable for a given application.

Freeze Protection for Backflow Preventers

Codes require backflow preventers to be installed outside. Depending on the climate, codes may specify taking steps to protect backflow preventers from freezing. Freeze protection is a good idea even if the codes don't require it. Install insulated protective cases around exposed backflow preventers and wrap pipes with heat tape. Always refer to the manufacturer's instructions first to ensure that you are using appropriate materials and methods.

3.1.0 Air Gaps

An air gap is a physical separation between a potable water supply line and the flood-level rim of a fixture (see *Figure 3*). It is the simplest form of backflow preventer. Air gaps can be used in low-hazard intermittent pressure applications with a risk of back siphonage. Ensure that the supply pipe terminates above the flood-level rim at a distance that is at least twice the diameter of the pipe. The air gap should never be less than 1 inch. Some codes have specific requirements for fixtures with openings that are less than 1 inch (see *Table 1*).

All fixture faucets must incorporate an air gap. Air gaps can be used to prevent back pressure, back siphonage, or both. Never submerge hoses or other devices connected to the potable water supply in a basin that may contain contaminated liquid. Even seemingly innocent activities such as filling up a wading pool with a garden hose could be a problem if, elsewhere, workers digging with a backhoe accidentally cut through the water main.

NOTE

Remember to calculate the pressure drop across backflow preventers when sizing water supply systems.

Containment and Isolation Backflow Preventers

Backflow preventers installed at the meter or curb protect the public water supply system. They are called containment backflow preventers. Backflow preventers on a house system protect the occupants from external contaminants. They are called isolation backflow preventers.

Backflow Preventers in Restaurants

Many fixtures and appliances in restaurants depend on clean, fresh water. Steamers, dishwashers, coffee machines, icemakers, and soda dispensers all connect to the fresh water supply. They all require backflow preventers. Public health inspectors regularly test these systems to ensure that they are working properly. A restaurant in violation of the local health code can be shut down. An accidental cross-connection, therefore, could not only cause sickness and death but could also damage a restaurant's reputation and threaten its economic survival.

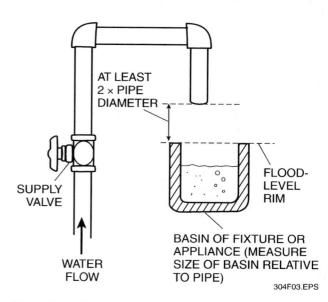

Figure 3 ◆ Air gap.

Table 1 Minimum Air Gaps From a Model Code

Fixture	Minimum Air Gap	
	Away from wall (inches)	Close to wall (inches)
Lavatories and other fixtures with an effective opening not greater than ½ inch in diameter	1	1½
Sinks, laundry trays, gooseneck back faucets, and other fixtures with effective openings not greater than ¾ inch in diameter	1½	2½
Over-rim bath fillers and other fixtures with effective openings not greater than 1 inch in diameter	2	3
Drinking water fountains, single orifice not greater than 7/16 inch in diameter or multiple orifices with a total area of the same area	1	1½
Effective openings greater than 1 inch	Two times the diameter of the effective opening	Three times the diameter of the effective opening

Because air gaps permit a loss of system pressure, they are most often used at the end of a service line where they empty into storage tanks or reservoirs. Never attempt to eliminate splashing during drainage by moving the end of the line into the receptacle. This defeats the purpose of the air gap and creates a cross-connection situation.

3.2.0 Atmospheric Vacuum Breakers

If it is not possible to install an air gap where there is a risk of back siphonage, consider using an **atmospheric vacuum breaker** (see *Figure 4*). Atmospheric vacuum breakers (AVBs) use a silicon disc float to control water flow. In normal operation, the float rises to allow potable water to enter. If back siphonage causes low pressure, air enters the breaker from vents and forces the disc into its seat. This shuts down the flow of fresh water (see *Figure 5*).

Use AVBs for intermittent-use, low-hazard nonpotable water systems where there is a risk of back siphonage, such as the following:

- Commercial dishwashers
- Sprinkler systems
- Outdoor faucets

Some codes allow AVBs to be used where a pipe or hose terminates below the flood-level rim of a basin, as long as no conditions for back pressure are present (see *Figure 6*). Because AVBs are meant to be used only in low-hazard applications, AVBs do not offer sufficient protection for health care or food processing applications. Other protection upstream of the AVB would be required in such

Figure 4 ◆ Cutaway of an atmospheric vacuum breaker.

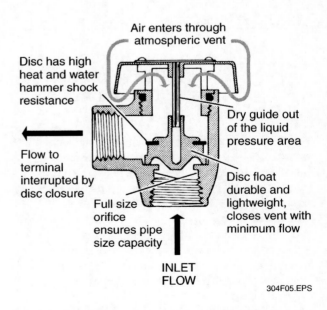

Figure 5 ◆ Atmospheric vacuum breaker in the closed position.

applications. Refer to your local applicable code before installing AVBs to ensure that they are permitted for such applications.

AVBs are designed to operate at normal air pressure. Do not use them where there is a risk of back pressure backflow. Check the manufacturer's specifications for the proper operating temperatures and pressures. If conditions exceed these limits, consult the manufacturer first. Unlike other types of backflow preventers, AVBs are not designed to be routinely tested.

Install an AVB after the last control valve, and ensure that it is at least 6 inches above the fixture's flood-level rim. Install the AVB so that the fresh water intake is at the bottom, and ensure that the water supply is flowing in the same direction as the arrow on the breaker. Use AVBs on applications with intermittent supply pressure. Do not install shutoff valves downstream from an AVB.

Hose connection vacuum breakers (see *Figure 7*) are designed for use where there is a hose connection. Some codes require them when there is a low-hazard risk of back siphonage in hoses attached to any of the following:

- Sill cocks (hose bibbs)
- Laundry tubs
- Service sinks
- Photograph developing sinks
- Dairy barns
- Wash racks

Hose connection vacuum breakers work the same way and under the same conditions as AVBs. Note that hose connections do not allow water to drain from an outdoor faucet that is not protected from freezing. If there is a danger of freezing, install a non-freeze faucet with a built-in vacuum breaker. This will allow the faucet to drain. Both types of AVBs can be equipped with strainers.

3.3.0 Pressure-Type Vacuum Breakers

Install **pressure-type vacuum breakers** (see *Figure 8*) to prevent back siphonage only, in either low- or high-hazard applications. They should be used in water lines that supply the following:

- Cooling towers
- Swimming pools
- Heat exchangers
- Degreasers
- Lawn sprinkler systems with downstream zone valves

A pressure-type vacuum breaker (PVB) has a spring-loaded check valve and a spring-loaded air inlet valve (see *Figure 9*). The PVB operates when line pressure drops to atmospheric pressure or less.

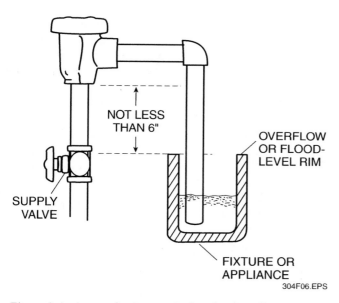

Figure 6 ◆ Atmospheric vacuum breaker installation.

Figure 7 ◆ Hose connection vacuum breaker.

Figure 8 ◆ Pressure-type vacuum breaker, ¾ inches through 2 inches.

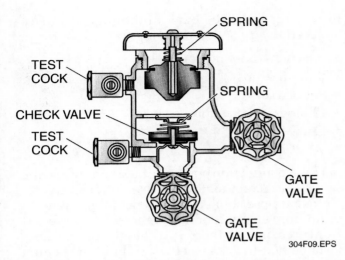

Figure 9 ♦ Cutaway illustration of a pressure-type vacuum breaker, ¾ inches through 2 inches.

The float valve opens the vent to the outside air. The check valve closes to prevent back siphonage.

PVBs are suitable for applications with constant inlet water pressure. PVBs are equipped with **test cocks**, which are valves that allow testing of the individual pressure zones within the device. They can also be equipped with strainers if required. Shutoff valves can be installed downstream from PVBs.

3.4.0 Dual-Check Valve Backflow Preventer Assemblies

Homes are full of potential sources of contamination. The most common include the following:

- Hose-attached garden spray bottles
- Lawn sprinkler systems
- Bathtub whirlpool adapters
- Hot tubs
- Water closet bowl deodorizers
- Wells or back-up water systems
- Photography darkrooms
- Misapplied pest extermination chemicals

To keep residential backflow from reaching the water main, use a **dual-check valve backflow preventer assembly** (DC) (see *Figure 10*). These preventers feature two spring-loaded check valves. They work similar to **double-check valve assemblies** (DCVs). Note that two check valves in series are not automatically a backflow preventer. A DC is a specially designed and sized single unit.

DCs protect against back pressure and back siphonage in low-hazard applications. Because DCs are smaller, they are commonly used in residential installations. These preventers tend to be less reliable than DCVs. Unlike DCVs, they are normally not fitted with shutoff valves or test cocks. This makes it difficult for plumbers to remove, test, or replace them. Check the local code before installing DCs. Install DCs on the customer side of a residential water meter. They can be installed either horizontally or vertically, depending on the available space and the existing piping arrangements.

3.5.0 Double-Check Valve Assemblies

Where heavy-duty protection is needed, install a double-check valve assembly to protect against back siphonage, back pressure, or both in low-hazard, continuous pressure applications (see *Figure 11*). A DCV has two spring-loaded check valves and includes a test cock. They are commonly used in food processing steam kettles and apartment buildings. During normal operation, both check valves are open; however, backflow causes the valves to seal tightly.

 WARNING!
Never install brass DCVs in soda carbonation systems. The carbon dioxide (CO_2) gas used to add bubbles to soda will chemically react with brass and create carbonic acid, a liquid that is lethal if ingested. Use stainless steel DCVs instead.

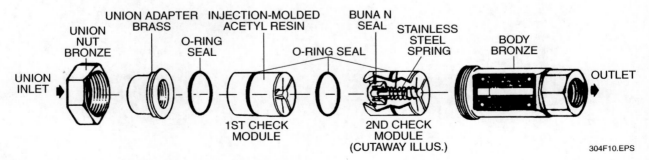

Figure 10 ♦ Dual-check valve backflow preventer assembly.

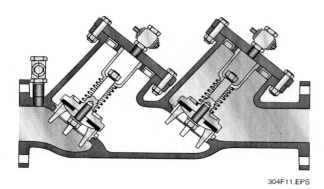

Figure 11 ♦ Double-check valve assembly.

DCVs are available in a wide range of sizes from ¾ inch up to 10 inches. DCVs have a gate or ball shutoff valve at each end. If required, DCV inlets can be equipped with strainers.

3.6.0 Reduced-Pressure Zone Principle Backflow Preventer Assemblies

Reduced-pressure zone principle backflow preventer assemblies (RPZs) protect the water supply from contamination by dangerous liquids (see *Figure 12*). They may be used in both low- and high-hazard applications where there is a risk of back siphonage and back pressure.

RPZs are the most sophisticated type of backflow preventer. They feature two spring-loaded check valves plus a hydraulic, spring-loaded pressure differential relief valve. The relief valve is located underneath the first check valve. The device creates a low pressure between the two check valves. If the pressure differential between the inlet and the space between the check valves drops below a specified level, the relief valve opens. This allows water to drain out of the space between the check valves (see *Figure 13*). The relief valve will also operate if one or both check valves fail. This offers added protection against backflow. RPZs also have shutoff valves at each end. Sometimes they are also fitted with strainers (refer to *Figure 12*).

Use an RPZ if a direct connection is subject to back siphonage, back pressure, or both. A wide variety of industrial and agricultural operations require RPZs, including:

- Car washes
- Poultry farms
- Dairies
- Medical and dental facilities
- Wastewater treatment plants
- Manufacturing plants

 WARNING!
Codes prohibit the installation of any type of bypass around a backflow preventer. Accidental backflow through the bypass will pollute or contaminate the system and can result in a serious health hazard.

Some codes require RPZs on additional types of installations. RPZs are available for all operating temperatures and pressures. Refer to your local code to see where RPZs are required. Many RPZs come with union or flanged connections, which allow fast and easy removal for servicing. Install RPZs so that their lowest point is at least 12 inches above grade, and provide support for the RPZ (see *Figure 14*). Otherwise, sagging could damage the RPZ. Select a safe location that will help prevent damage to and vandalism of the RPZ. Install a **manufactured air gap** on all RPZ installations (see *Figure 15*). It acts as a drain receptor to prevent siphonage.

Some codes permit two RPZs to be installed in a parallel fashion (see *Figure 16*). This allows the system to continue working if an RPZ malfunctions or is being serviced. If two RPZs are used in parallel, a smaller size can be used than if only one RPZ were installed. Locate a strainer in the pipe before both backflow preventers. If there is sediment in the system, it may clog a single strainer. In that case, install separate strainers in advance of each RPZ. Consult with the project engineer and the local plumbing code official before installing parallel backflow preventers.

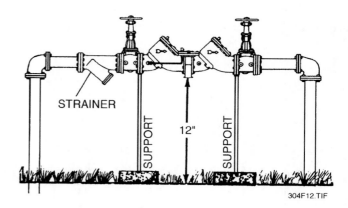

Figure 12 ♦ Installation of a reduced-pressure zone principle backflow preventer assembly.

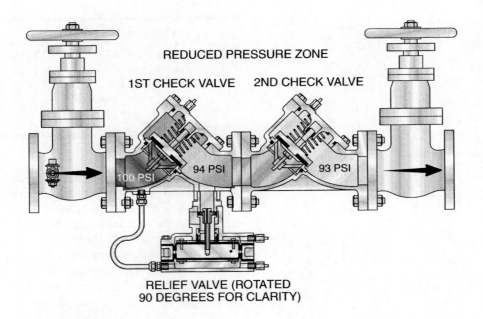

Figure 13 ♦ Operation of a reduced-pressure zone principle backflow preventer assembly with check valves parallel to flow.

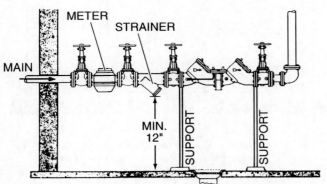

Figure 14 ♦ Reduced-pressure zone principle backflow preventer assembly installed with supports.

Figure 15 ♦ Manufactured air gap.

DID YOU KNOW?
Cross-Connection and the Spread of Bacteria

E. coli and Legionnaire's disease, two dangers spread by cross-connection bacteria, are safe when they stay where they belong. When they get into the water supply, they can be deadly. Escherichia coliform, also known as E. coli, is a common type of bacteria. It can be found in animal and human intestines, where it provides natural vitamins, but E. coli in animal wastes can be dangerous. It can enter the fresh water supply through untreated runoff and cross-connections. E. coli turns safe potable water into a disease delivery system and can contaminate meat and other foods. Because plumbers are responsible for installing and maintaining safe plumbing systems, E. coli is a real concern for them, too.

Health officials regularly test for E. coli outbreaks in the water supply. When one occurs, it can cause chaos and severe illness. Residents have to use bottled water or tanker trucks have to bring in water for drinking and cooking. Contaminated water must be boiled before people can use it to even brush their teeth. Testing, sampling, and cleanups can take weeks. Sometimes months go by before the health department declares a system completely clean.

Bacteria causes other health problems, including Legionnaire's disease. The first major reported outbreak of this disease occurred at an American Legion conference in 1976. Scientists named the bacteria Legionella. It reproduces in warm water. Legionella can grow in hot water tanks and some plumbing pipes. People inhale the bacteria through contaminated water mist from large air conditioning installations, whirlpool spas, and showers. It is not contagious, but between 5 and 30 percent of people who contract Legionnaire's disease die from it. Antibiotics do not kill Legionella, but they can keep the bacteria from multiplying.

Plumbers can help block the spread of disease caused by waterborne bacteria. Design plumbing systems so that cross-connections can't happen. Install interceptors and pumps to keep out harmful waste and to prevent water from stagnating. Plumbers play a vital role in maintaining public health. Doctors can treat disease, but plumbers can help prevent it.

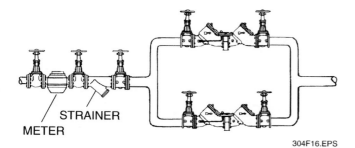

Figure 16 ◆ Reduced-pressure zone principle backflow preventer assemblies installed in parallel.

4.0.0 ◆ SPECIALTY BACKFLOW PREVENTERS

Often, low-flow and small supply lines require backflow preventers. Use a backflow preventer with an **intermediate atmospheric vent** for this kind of line (see *Figure 17*). The intermediate atmospheric vent is a form of DCV that also protects against back pressure and back siphonage in low-hazard applications. This type of breaker is therefore suitable for use on the following installations:

- Most laboratory equipment
- Processing tanks
- Sterilizers
- Residential boiler feeds

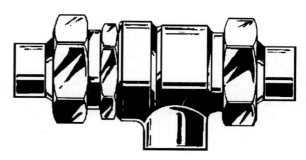
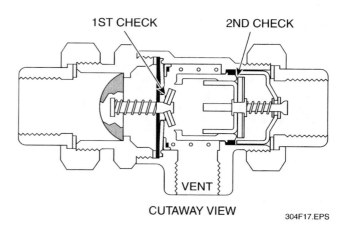

Figure 17 ◆ Intermediate atmospheric vent vacuum breaker.

ON THE LEVEL

Selecting Backflow Preventers

When determining which type of backflow preventer to install, you must always consider its application. Because backflow preventers help to maintain public health, liability risks affect how potential hazards are defined. The type of installation varies depending on the degree of hazard presented by the potential contaminants. Therefore, always check your local code before installing a backflow preventer.

Your local code or health department will identify low-hazard and high-hazard applications and specify the types of backflow preventers that can be used in such cases. Because plumbers are responsible for purchasing and installing backflow preventers, you should take the time to become familiar with the brands approved for use in your area.

Laboratories and hospitals use chemicals that can cause illness or death. Special backflow preventers help protect the fresh water supply in such applications. Laboratory faucets with hose attachments require **in-line vacuum breakers** (see *Figure 18*). In-line vacuum breakers work the same way as atmospheric vacuum breakers. A disc allows fresh water to flow to the faucet. Negative pressure causes the disc to close the water inlet. In-line vacuum breakers protect against back siphonage in high-hazard applications. Install them on new and existing faucets. Changes to the plumbing are not required.

WARNING!
Never install a backflow preventer in a pit. Refer to your local code for approved installation options.

5.0.0 ♦ BACKFLOW PREVENTER TESTING

CAUTION
Under no circumstances should an uncertified person attempt to test backflow preventers. Likewise, a person who is certified to test backflow preventers is not necessarily certified to also repair water supply systems. Plumbers can use a certified tester's report to determine repair needs, however.

Backflow preventers can break down. When they do, they need to be repaired or replaced. Many state codes require regular testing of preventers. Only certified technicians can service and test backflow preventers. Certified testers are licensed in each state.

Testers who are not plumbers are not allowed to repair or replace backflow preventers. However, a plumber may work on backflow preventers and then have a certified tester test the device. The tester will then send a report of the repair work to the authority having jurisdiction.

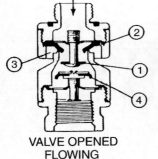

VALVE OPENED FLOWING UNDER PRESSURE

With flow through valve, primary check (1) opens away from diaphragm seal (2). Atmospheric port (3) remains closed by deflection of diaphragm seal. Secondary check (4) opens away from downstream seat, permitting flow of water through valve.

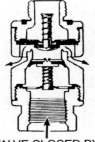

VALVE CLOSED BY BACK-SIPHONAGE IN SYSTEM

With a back-siphonage condition, secondary check seals tightly against downstream seat. Primary check seals against diaphragm. Atmospheric port is open, permitting air to enter air break chamber. In the event of fouling of downstream check valve, leakage would be vented to the atmosphere through the vent port.

304F18.EPS

Figure 18 ♦ In-line vacuum breaker.

ON THE LEVEL

Certified Backflow Preventer Assembly Testers

In many states, plumbers can install backflow preventers. However, preventer testing must be left to certified testers. Codes provide stiff penalties for a noncertified plumber who carries out an inspection. Refer to the local code.

Testers must take written and practical exams offered at a site authorized by the local chapter of the American Water Works Association (AWWA). Applicants must have a high school diploma or General Education Development (GED) equivalent, be at least 18 years old, and speak and understand English. In many states, testers do not have to be licensed plumbers. The certification is good for three years. Never rush to save time; you might end up wasting even more time as a result!

Many backflow preventers have test cocks, which are used for testing (see *Figure 19*). Foreign matter blocking the valves can cause a malfunction, so many backflow preventers have strainers to trap dirt and other matter. Strainers can be cleaned without having to disconnect the preventer. If sediment interferes with the operation of a backflow preventer, the system needs to be flushed.

NOTE

Backflow preventer test equipment must be calibrated yearly. Refer to the manufacturer's instructions for the calibration schedule and method.

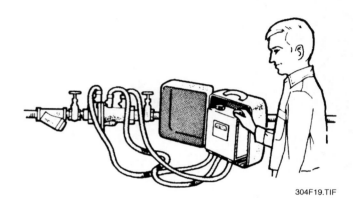

Figure 19 ◆ Testing pressure zones.

Review Questions

Sections 3.0.0–5.0.0

1. All of the following are factors in selecting the most appropriate backflow preventer for an installation *except*:
   a. Type of backflow
   b. Piping material
   c. Risk of cross-connections
   d. Health risk posed by contaminants

2. An air gap should never be less than _____ inch(es).
   a. ¼
   b. ½
   c. 1
   d. 2

3. _____ only protect against back siphonage.
   a. Atmospheric vacuum breakers
   b. Air gaps
   c. Double-check valve assemblies
   d. Reduced-pressure zone principle backflow preventers

4. Protect the house side of a residential water meter by installing a _____.
   a. hose connection vacuum breaker
   b. dual-check valve backflow preventer
   c. double-check valve assemblies
   d. reduced-pressure zone principle backflow preventer

5. The RPZ type has a _____ that works with the two spring-loaded check valves to provide strong protection against backflow.
   a. test cock
   b. spring-loaded air inlet valve
   c. hose connection vacuum breaker
   d. spring-loaded pressure differential relief valve

In Questions 6–11, specify the approved hazard levels for the following backflow preventers according to the *2003 IPC*.

6. Air gap: _____

7. Atmospheric vacuum breaker: _____

8. Pressure-type vacuum breaker: _____

9. Dual-check valve backflow preventer assembly: _____

10. Double-check valve assembly: _____

11. Reduced-pressure zone principle backflow preventer assembly: _____

12. Use an intermediate atmospheric vent backflow preventer on _____ supply lines.
    a. high-pressure
    b. large
    c. small
    d. contaminant

13. Laboratory faucets with hose attachments require _____.
    a. pressure-type vacuum breakers
    b. in-line vacuum breakers
    c. reduced-pressure zone principle backflow preventer assemblies
    d. dual-check valve backflow preventer assemblies

14. A certified backflow preventer tester does not have to be a plumber.
    a. True
    b. False

15. Many backflow preventers have _____, which are used for testing.
    a. check valves
    b. test cocks
    c. pressure gauges
    d. relief valves

Summary

Problems and malfunctions in a plumbing system can force water to flow backward through a system. Backward flow can draw contaminants into a plumbing system through a cross-connection. This is called backflow. Back pressure and back siphonage are the two kinds of backflow. Backflow preventers protect the potable water supply from contamination. There are six main types of backflow preventers: air gap, atmospheric vacuum breaker, pressure-type vacuum breaker, double-check valve, dual-check valve assembly, and reduced-pressure zone principle assembly. Each of these is designed for specific applications, and some work on both types of backflow.

Codes require certified technicians to service and test backflow preventers. Do not attempt to service or test a preventer without certification. Regular tests ensure that backflow preventers continue to operate correctly. Always consult with the project engineer and local code officials before selecting and installing a backflow preventer. Public safety depends on the reliable operation of backflow preventers, and the law requires it.

Notes

Trade Terms Introduced in This Module

Atmospheric vacuum breaker: A backflow preventer designed to prevent back siphonage by allowing air pressure to force a silicon disc into its seat and block the flow.

Back siphonage: A form of backflow caused by subatmospheric pressure in the water system, which results in siphoning of water from downstream.

Backflow: An undesirable condition that results when nonpotable liquids enter the potable water supply by reverse flow through a cross-connection.

Backflow preventer with intermediate atmospheric vent: A specialty backflow preventer used on low-flow and small supply lines to prevent back siphonage and back pressure. It is a form of double-check valve.

Cross-connection: A direct link between the potable water supply and water of questionable quality.

Double-check valve assembly (DCV): A backflow preventer that prevents back pressure and back siphonage through the use of two spring-loaded check valves which seal tightly in the event of backflow. DCVs are larger than dual-check valve backflow preventers and are used for heavy-duty protection. DCVs have test cocks.

Dual-check valve backflow preventer assembly (DC): A backflow preventer that uses two spring-loaded check valves to prevent back siphonage and back pressure. DCs are smaller than double-check valve assemblies and are used in residential installations. DCs do not have test cocks.

High hazard: A classification denoting the potential for an impairment of the quality of the potable water which creates an actual hazard to the public health through poisoning or through the spread of disease by sewage, industrial fluids, or waste. This type of water quality impairment is called contamination.

Hose connection vacuum breaker: A backflow preventer designed to be used outdoors on hose bibbs to prevent back siphonage.

In-line vacuum breaker: A specialty backflow preventer that uses a disc to block flow when subjected to back siphonage. It works the same as an atmospheric vacuum breaker.

Low hazard: A classification denoting the potential for an impairment in the quality of the potable water to a degree which does not create a hazard to the public health but which does adversely and unreasonably affect the aesthetic qualities of such potable water for domestic use. This type of water quality impairment is called pollution.

Manufactured air gap: An air gap that can be installed on a reduced-pressure zone principle backflow preventer to prevent back pressure and back siphonage.

Pressure-type vacuum breaker: A backflow preventer installed in a water line that uses a spring-loaded check valve and a spring-loaded air inlet valve to prevent back siphonage.

Reduced-pressure zone principle backflow preventer assembly: A backflow preventer that uses two spring-loaded check valves plus a hydraulic, spring-loaded pressure differential relief valve to prevent back pressure and back siphonage.

Test cock: A valve in a backflow preventer that permits the testing of individual pressure zones.

Resources & Acknowledgments

Additional Resources

This module is intended to be a thorough resource for task training. The following reference works are suggested for further study. These are optional materials for continued education rather than for task training.

Backflow Prevention: Theory and Practice, 1990. Robin L. Ritland. Dubuque, IA: Kendall/Hunt Publishing Company.

Manual of Cross-Connection Control, Ninth Edition, 1993. Foundation for Cross-Connection Control and Hydraulic Research. Los Angeles, CA: University of Southern California.

Recommended Practice for Backflow Prevention and Cross-Connection Control, Second Edition, Manual M14, 1990. Denver, CO: American Water Works Association.

Figure Credits

International Code Council, Inc., *2003 International Plumbing Code*, 304SA01, Table 1

U.S. EPA Office of Ground Water and Drinking Water, 304F02, 304F09, 304F11, 304F13A, 304F13B, 304F17B

Watts Regulator Company, 304F01, 304F03–304F08, 304F10, 304F12, 304F14–304F16, 304F17A, 304F18, 304F19

References

"50 Cross-Connection Questions, Answers, and Illustrations Relating to Backflow Prevention Products and Protection of Safe Drinking Water Supply," 2001. Publication F-50, Watts Regulator Company, www.wattsreg.com/pdf-files/brochures/F-50.pdf.

2003 International Plumbing Code. Falls Church, VA: International Code Council.

Dictionary of Architecture and Construction, Third Edition. 2000. Cyril M. Harris, ed. New York: McGraw-Hill.

Cross-Connection Control Manual, 2003. Publication 816-R-03-002, United States Environmental Protection Agency, Office of Water (4606M), Washington, DC.

Pipefitters Handbook. 1967. Forrest R. Lindsey. New York: Industrial Press Inc.

"Protect Your Water Supply from Agricultural Chemical Backflow," 1993. Extension Bulletin E-2349, Robert H. Wilkinson and Julie Stachecki, Michigan State University Cooperative Extension Service, East Lansing, MI, http://www.ag.uiuc.edu/~vista/html_pubs/back/back.htm.

"Questions and Answers About Cross-Connection Control," 2001. Florida Department of Environmental Protection, www.mindspring.com/~loben/dep-info.htm.

CONTREN® LEARNING SERIES — USER UPDATE

The NCCER makes every effort to keep these textbooks up-to-date and free of technical errors. We appreciate your help in this process. If you have an idea for improving this textbook, or if you find an error, a typographical mistake, or an inaccuracy in NCCER's Contren® textbooks, please write us, using this form or a photocopy. Be sure to include the exact module number, page number, a detailed description, and the correction, if applicable. Your input will be brought to the attention of the Technical Review Committee. Thank you for your assistance.

Instructors – If you found that additional materials were necessary in order to teach this module effectively, please let us know so that we may include them in the Equipment/Materials list in the Annotated Instructor's Guide.

Write: Product Development and Revision
National Center for Construction Education and Research
P.O. Box 141104, Gainesville, FL 32614-1104

Fax: 352-334-0932

E-mail: curriculum@nccer.org

Craft _____ Module Name _____

Copyright Date _____ Module Number _____ Page Number(s) _____

Description
_____
_____
_____
_____

(Optional) Correction
_____
_____
_____

(Optional) Your Name and Address
_____
_____
_____

Plumbing Level Three

02305-06
Types of Venting

02305-06
Types of Venting

Topics to be presented in this module include:

1.0.0	Introduction	.5.2
2.0.0	How Vents Work	.5.2
3.0.0	Designing a Vent Installation	.5.2
4.0.0	Types of Vents	.5.10

Overview

Without vents, plumbing systems would not work. Vents allow air to enter the system when wastewater flows out of drain pipes. They prevent trap seals from draining by maintaining equalized air pressure throughout the drainage system.

Plumbers must custom design every vent system according to their local applicable code to provide the most efficient venting of all the fixtures in a building. Most systems have the same basic components. The stack vent, an extension of the main soil and waste stack, ends in a vent terminal through the roof. The building's central vent, the vent stack or main vent, begins at the base of the main stack and connects with the stack vent. Other vents connect to either the stack vent or the main vent. It is also very important to grade the venting system properly.

Plumbers can choose from many different types of vents to create the best DWV system possible. Individual vents effectively provide correct air pressure at a fixture's trap. With common vents, two fixtures share a single vent. Battery vents, usually used in commercial buildings, connect a series of fixtures to the vent stack. Wet vents, popular for venting residential bathroom fixtures with relatively low flow rates, also carry liquid waste. Air admittance vents, which operate like one-way valves, ventilate the DWV stack using the building's own air. Relief vents allow excess air pressure in the drainage system to escape. Finally, the innovative Sovent® system combines vent and waste stacks in one stack.

Focus Statement
The goal of the plumber is to protect the health, safety, and comfort of the nation job by job.

Code Note
Codes vary among jurisdictions. Because of the variations in code, consult the applicable code whenever regulations are in question. Referring to an incorrect set of codes can cause as much trouble as failing to reference codes altogether. Obtain, review, and familiarize yourself with your local adopted code.

Portions of this publication reproduce tables and figures from the *2003 International Plumbing Code* and the *2000 IPC Commentary*, International Code Council, Inc., Falls Church, Virginia. Reproduced with permission. All rights reserved.

Objectives

When you have completed this module, you will be able to do the following:

1. Describe the scientific principles of venting.
2. Design vent systems according to local code requirements.
3. Sketch the different types of vents.
4. Construct given vent configurations.
5. Install the different types of vents correctly.
6. Select correct fittings for vents.

Trade Terms

Air admittance vent
Back pressure
Back vent
Branch interval
Branch vent
Circuit vent
Combination waste and vent system
Common vent
Continuous vent
Flat vent
Hydraulic gradient
Indirect or momentum siphonage
Individual vent
Loop vent
Relief vent
Re-vent
Self-siphonage
Sovent® system
Stack vent
Vent
Vent stack
Vent terminal
Wet vent
Yoke vent

Required Trainee Materials

1. Appropriate personal protective equipment
2. Sharpened pencils and paper
3. Copy of local applicable code
4. Calculator

Prerequisites

Before you begin this module, it is recommended that you successfully complete *Core Curriculum; Plumbing Level One; Plumbing Level Two; Plumbing Level Three*, Modules 02301-06 through 02304-06.

This course map shows all of the modules in the third level of the *Plumbing* curriculum. The suggested training order begins at the bottom and proceeds up. Skill levels increase as you advance on the course map. The local Training Program Sponsor may adjust the training order.

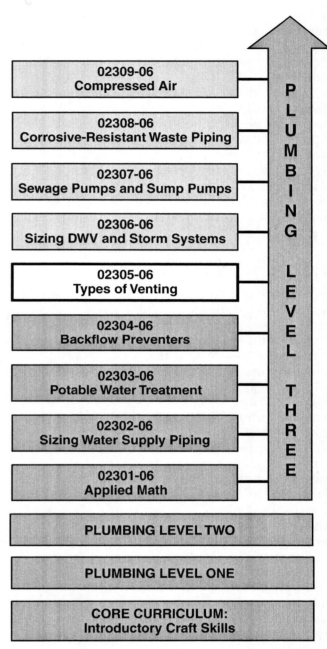

1.0.0 ♦ INTRODUCTION

Vents allow air to enter and exit a drain, waste, and vent (DWV) stack and protect the trap seal from siphoning. Vents work in conjunction with drains to enable wastes to flow from fixtures into the waste stack, where they are carried into the sewer. Without vents, plumbing systems would not work. Odors and contaminants would build up, and people would become sick. With proper venting, wastes are carried away efficiently and effectively.

Earlier in the *Plumbing* curriculum you were introduced to vents as components of DWV systems. In this module you will learn how vents operate. You will also review the different types of vents that can be installed in a DWV system. Many types of vents can be grouped together into broad categories. This will make it easier for you to remember what they are called and how they are used. Sizing vents will be covered in more detail in the module *Sizing DWV and Storm Systems*.

2.0.0 ♦ HOW VENTS WORK

Vents prevent trap seals from draining when water flows through the DWV system. They do this by maintaining equalized air pressure throughout the drainage system. Air is used to equalize pressure that is exerted when wastewater drains from a fixture into the system.

You can visualize the principle of air pressure by observing the behavior of water in a straw. If you fill a straw with water and put your thumb over the top, the water remains in the straw. As soon as you remove your thumb, the water drains out. This is because water cannot flow out of the straw unless air can flow into the straw at the same rate, filling the space left by the draining water. In other words, by replacing the water in the straw with an equivalent amount of air, an equal pressure is maintained inside the straw.

To achieve the same effect in a DWV installation, plumbers install vents. Vents allow air to enter the system when wastewater flows out of the drain pipe. They ventilate the drainage system to the outside air and also keep the pipes from clogging and siphoning the trap dry. Air pressure, therefore, is important for creating and maintaining proper flow in drainage pipes. Also, normal air pressure throughout the DWV system helps maintain the water seal in fixture traps. System pressure should be equal to air pressure.

Too much or too little air pressure in the vent pipes can cause the DWV system to malfunction. Low pressure in the waste piping (also called a partial vacuum or negative pressure) can suck the water seal out of a trap and into the drain. This is called **self-siphonage** (see *Figure 1*). When the discharge from another drain creates lower pressure in the system, it can draw a trap seal into the waste pipe (see *Figure 2*). This is called **indirect or momentum siphonage**. Properly placed vents prevent both types of siphonage. To further protect against indirect siphonage, ensure that the stack is properly sized.

Likewise, excess pressure in vent piping can have the reverse effect, blowing the trap seal out through the fixture. Air in the stack that is compressed by the weight of water above can force the water out of a trap as the slug of air passes a fixture. This phenomenon, called **back pressure**, has a tendency to happen in taller buildings (see *Figure 3*). Use a vent near the fixture traps or at the point where the piping changes direction to prevent back pressure. In some cases, downdrafts of air entering a vent on a roof can create a condition similar to back pressure. Install roof vents away from roof ridges and valleys that stir up downdrafts.

3.0.0 ♦ DESIGNING A VENT INSTALLATION

A vent system must provide adequate venting for all the fixtures in a building. Each building has its own special requirements, which are dictated by the building's location, use, and level of occupancy.

Vent systems must also be specially designed. However, most vent systems, regardless of where and how they are installed, have the same basic components. Plus, vents can be installed using procedures that are already familiar to you.

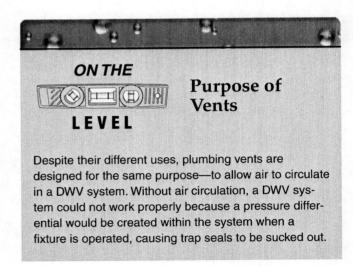

ON THE LEVEL

Purpose of Vents

Despite their different uses, plumbing vents are designed for the same purpose—to allow air to circulate in a DWV system. Without air circulation, a DWV system could not work properly because a pressure differential would be created within the system when a fixture is operated, causing trap seals to be sucked out.

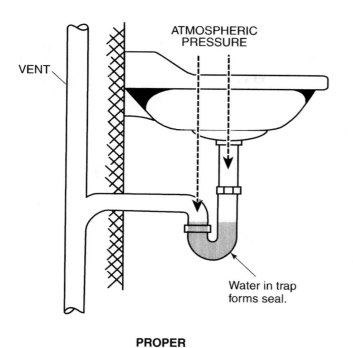
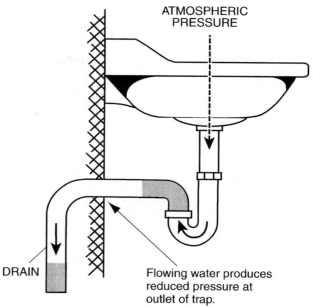

Figure 1 ♦ Self-siphonage resulting from improper venting.

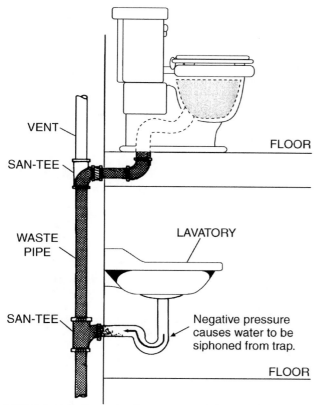

NOTE: Installation is for the purpose of demonstration only.

Figure 2 ♦ Indirect or momentum siphonage.

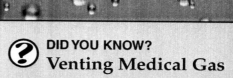

DID YOU KNOW?
Venting Medical Gas

Medical gas systems affect the treatment and care of hospital patients. Venting medical gas is a complicated task that requires additional training. To be able to complete this work, you must be at least a journeyman and have a medical gas certification. If you are not properly trained and certified, do not attempt to do this work. Seek the assistance of a trained professional.

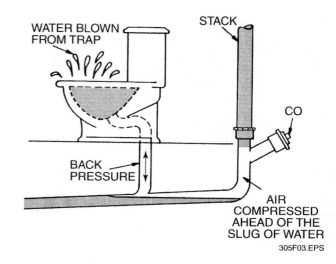

Figure 3 ♦ Back pressure.

MODULE 02305-06 ♦ TYPES OF VENTING   5.3

DID YOU KNOW?
The Discovery of Air Pressure

In fifteenth-century Italy, the Grand Duke of Tuscany ordered his engineers to build a giant pump. The pump was designed to raise water more than 40 feet above the ground. It worked by drawing water into a tall column by creating a partial vacuum. However, the engineers found that the pump could not raise water higher than 32 feet. The Grand Duke asked Evangelista Torricelli to solve the mystery. Torricelli (1608–1647) was a famous mathematician and a physicist.

Torricelli discovered that the pump worked fine; something else was preventing the water from reaching a higher level. To find the answer, he conducted a simple experiment. He filled a 3-foot long test tube with liquid mercury. Then he quickly turned the tube upside down into a bowl full of more mercury, making sure the mouth of the tube was below the surface of the mercury in the bowl.

Some of the mercury in the tube flowed into the bowl, but most stayed inside the upside-down tube. Between the column of mercury in the tube and the bottom—now the top—of the tube, there was a gap, a vacuum like the one created in the duke's pump.

Torricelli reasoned that the air in the atmosphere was pushing down on the mercury in the bowl, forcing some of the mercury to stay in the tube. Torricelli's pushing force is what we call air pressure. The same thing was happening to the duke's pump. The water could only rise to 32 feet because the weight of the surrounding air could not push it any higher.

We now know that air pressure at sea level exerts a pressure of approximately 14.7 pounds per square inch (psi). In other words, at any given moment you are being gently squeezed on all sides by the weight of the air around you. Air pressure affects things that most people may never think about: how long it takes water to boil, how birds fly, or how a pump works.

3.1.0 Components of a Vent System

The relationship among the vents in a DWV system is illustrated in *Figure 4*. The main soil and waste stack runs between the building drain and the highest horizontal drain in the system. Above the highest horizontal drain is the **stack vent**, which is an extension of the main soil and waste stack. The stack vent allows air to enter and exit the plumbing system through the roof. The stack vent also serves as the **vent terminal** for other vent pipes.

The central vent in a building is called the **vent stack**. The vent stack permits air circulation between the drainage system and the outside air. Other vents may be connected to it. The vent stack, which is also called the main vent, runs vertically. It is usually located within a few feet of the main soil and waste stack. The vent stack begins at the base of the main stack and continues until it connects with the stack vent. It may also extend through the roof on an independent path. Branches connect one or more individual fixture vents to the vent stack. Codes require that vents go out through the roof unobstructed.

Vents that run horizontally to a drain are called **flat vents**. Flat vents should not be used in floor-mounted fixtures. They can cause waste to back up into the vent and block the fixture drain (see *Figure 5*). Codes prohibit vents from being turned from vertical to horizontal until at least 6 inches above the flood level rim of the fixture.

ON THE LEVEL
Vent Terminals Through the Roof

In the DWV system shown in *Figure 4*, the designer could also extend the right-hand vent stack through the roof with its own vent terminal (VTR). Both options are permitted by code. Remember, a vent must go through the roof unobstructed.

NOTE
Remember, all vents need to go to the outside unobstructed, according to code.

The vent terminal is the point at which a vent terminates. Fixture vents terminate at the point where they connect to the vent stack or directly to the stack vent. Stack vents and some vent stacks terminate above the roof. Local codes govern the location, size, and type of roof vents (see *Figure 6*).

5.4 PLUMBING ◆ LEVEL THREE

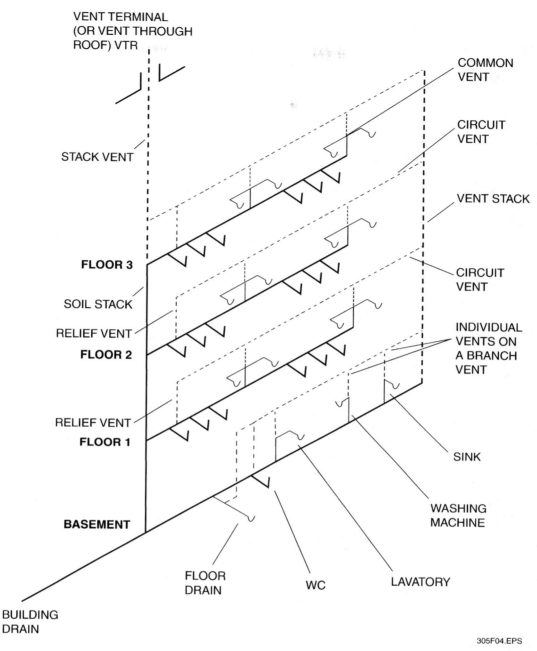

Figure 4 ◆ Vents in a DWV system.

Follow the code closely when selecting and installing a roof vent terminal. When installing a roof vent, consider the following issues:

- Code requirements for freeze protection
- The use of permitted prefabricated vent flashing
- Pitched vs. flat roof
- Code requirements for flashing collars

When vent terminals are installed, they should be protected from being plugged by snow accumulation or other means. Some codes require that the dimensions of the vent terminal be increased before exiting the roof. Your local code will provide specific information on vent terminals that are permitted for use in your area. Your code will also include dimensions for property lines, air intakes for heating, ventilating, and air conditioning (HVAC) systems, and air compressor intakes.

3.2.0 Vent Grades

Grade is the slope or fall of a line of pipe in reference to a horizontal plane. In a DWV system, grade allows solid and liquid wastes to flow out under the force of gravity.

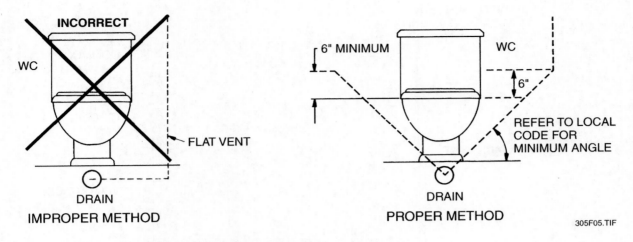

Figure 5 ◆ Improper and proper vent installation for a floor-mounted fixture.

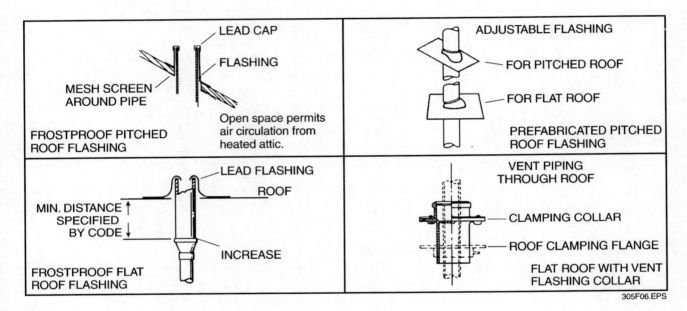

Figure 6 ◆ Types of vent terminals.

Horizontal vent pipes also need to be installed with proper grade. This will allow condensation to drain without blocking the vent pipe. Your local plumbing code will specify the degree of slope required for a horizontal vent. Most codes require that the vent be taken off above the soil stack centerline. Then it must rise at an angle, typically not less than 45 degrees from horizontal, to a point 6 inches above the fixture's flood level rim (see *Figure 7*). That way, if the drain pipe becomes blocked, solids cannot enter and potentially clog the vent instead.

The maximum grade between the trap weir and the vent pipe opening should not exceed one pipe diameter (see *Figure 8*). This grade is called the **hydraulic gradient**. The gradient determines the distance from the weir to the vent. If the vent is placed too far from the fixture trap, the vent opening will end up below the weir. This could cause self-siphoning. Codes include tables that specify the maximum distance the weir can be positioned from the vent based on pipe fall and diameter.

 WARNING!
Sewer gas contains methane. Methane can build up to explosive levels of concentration if a DWV system is vented improperly. Review your local code and all building plans to ensure that the system design has adequate venting.

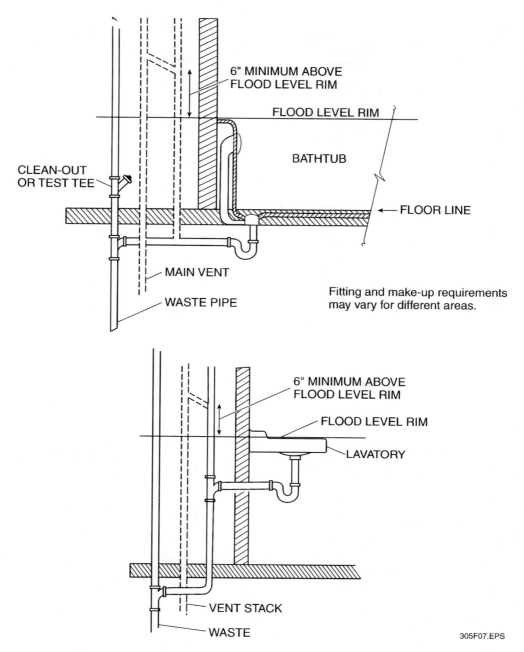

Figure 7 ◆ Connection to the vent stack.

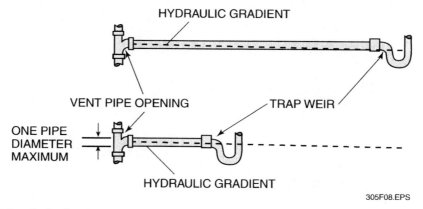

Figure 8 ◆ Determining the hydraulic gradient.

MODULE 02305-06 ◆ TYPES OF VENTING 5.7

Crown Vent

A vent opening that is located within two pipe diameters of a trap is called a crown vent (see illustration). **Crown vents are prohibited by code.** Relocate such vents further away from the trap. Never rush to save time; you might end up wasting even more time as a result!

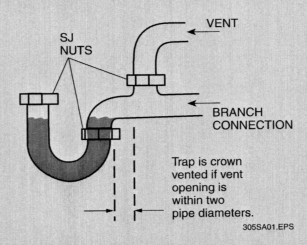

 DID YOU KNOW?
Plumbing Resources on the World Wide Web

You can find a lot of useful information on the Internet. It can be accessed easily and quickly from home and the office. Thanks to developments in wireless technology, it can also be accessed from the job site. Web sources are constantly updated to reflect changes in the industry and the profession. Here is a small sampling of websites that contain valuable information for plumbers:

Code Organizations
International Association of Plumbing & Mechanical Officials (IAPMO)—*www.iapmo.org*
International Code Council (ICC)—*www.iccsafe.org*
Plumbing-Heating-Cooling Contractors—National Association (PHCC)—*www.phccweb.org*

Standards and Testing
National Association of Home Builders Research Center—*www.nahbrc.org*
American Society of Sanitary Engineering—*www.asse-plumbing.org*

Books and Online Magazines
Amazon.com—*www.amazon.com*
Pipefitter.com—*www.pipefitter.com*
Plumbing & Mechanical Magazine—*www.pmmag.com*

Reference and Directories
PlumbingNet—*www.plumbingnet.com*
PlumbingWeb—*www.plumbingweb.com*
ThePlumber.com—*www.theplumber.com*

Also, be sure to bookmark the website of your local authority having jurisdiction. It will contain useful information. Remember, just because a site is on the Internet doesn't mean that it is accurate. Find out if an author or an organization is an established authority on the subject. With information, just like with anything in plumbing, it is better to be safe than sorry.

Review Questions

Sections 1.0.0–3.0.0

1. Vents allow _____ to equalize the pressure in a DWV system.
   a. waste
   b. water
   c. air
   d. solids

2. Self-siphonage occurs when _____ pressure in the waste piping sucks a water seal into a drain.
   a. back
   b. low
   c. high
   d. excess

3. Back pressure can be prevented by installing a vent in one of two places: near the fixture traps or _____.
   a. at the connection to the stack vent
   b. at the connection to the vent stack
   c. where the stack terminates
   d. where the piping changes direction

4. Each of the following is a consideration when designing vents for a DWV system for a particular building *except* _____.
   a. location
   b. level of occupancy
   c. pipe materials
   d. use

5. The _____ is located immediately above the highest horizontal drain in a DWV system.
   a. stack vent
   b. vent stack
   c. vent terminal
   d. crown vent

6. The vent stack begins _____ and continues until it connects with the stack vent.
   a. at the lowest point in the building
   b. adjacent to the main stack
   c. at the top of the main stack
   d. at the base of the main stack

7. Codes require that vents go out through the roof _____.
   a. capped
   b. unobstructed
   c. 12 inches
   d. through a structural member

8. The hydraulic gradient is the maximum grade between a trap weir and the _____.
   a. nearest flat vent
   b. vent pipe opening
   c. drain pipe opening
   d. nearest horizontal vent

9. Most codes require that horizontal vents rise at an angle, typically not less than _____ degrees from horizontal, to a point _____ inches above the fixture's flood level rim.
   a. 30; 6
   b. 30; 12
   c. 45; 6
   d. 45; 12

10. Most codes require that the vent be taken off below the soil stack centerline.
    a. True
    b. False

4.0.0 ♦ TYPES OF VENTS

Every venting system must be custom designed according to your local applicable code to provide the most efficient venting, taking into account the building's function, location, and physical layout. There are many different types of vents from which to choose. This gives plumbers the flexibility to create the best DWV system possible. Each vent is constructed to serve a particular purpose. Always refer to your local code to identify the vents that are approved for use in each individual application.

Vents are used in a variety of combinations with each other. Many vents have more than one name, depending on the code being used. To make it easier to remember all the different types and names of vents, it will help to begin by grouping them together into a few broad categories.

4.1.0 Individual Vents

A vent that runs from a single fixture directly to the vent stack is called an **individual vent**. Plumbers install, or **re-vent**, individual vents to provide correct air pressure at each trap. Because of this, they are perhaps the most effective means of venting a fixture. A continuous vent is a type of individual vent. It is a vertical extension of the drain to which it is connected. Install **continuous vents** far enough away from fixture traps to avoid crown venting. Connect continuous vents to the stack at least 6 inches above the fixture's flood level rim or overflow line.

A **back vent**, as its name suggests, is an individual vent installed at the back of a fixture. Use back vents to connect the fixture's drain pipe to the vent stack, or terminate back vents in the open air. In most cases, back vents should be connected to the drain pipe as close to the trap as possible. Back vents may incorporate a continuous vent (see *Figure 9*). Appropriate connection points, slope, and pipe sizes for back vents are spelled out in the building plans or your local code. A **branch vent** is any vent that connects a single fixture or a group of fixtures to a vent stack at least 6 inches above the flood level rim of the fixture.

4.2.0 Common Vents

Use a **common vent** where two similar fixtures are installed with all three of the following characteristics:

- The fixtures are back-to-back.
- The fixtures are on opposite sides of a wall.
- The fixtures are at the same height.

Common vents are also called unit vents. In a common vent, both fixture traps are connected to the drain with a short pattern sanitary cross (see *Figure 10*). The two fixtures share a single vent. The vent extends above the drain from the top connection of the sanitary cross. Common vents are used in apartments or hotels where the design calls for back-to-back fixtures and shared vents.

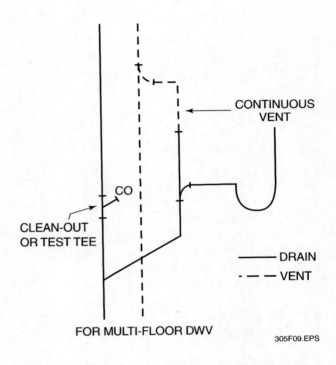

Figure 9 ♦ Back vent incorporating a continuous vent.

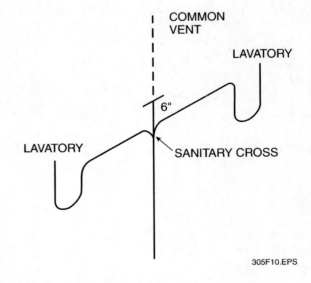

Figure 10 ♦ Common or unit vent.

4.3.0 Battery Vents

Many commercial buildings have large numbers of fixtures. It may be impractical or even impossible to provide each fixture with an individual vent. Horizontal vents connect a series, or battery, of fixtures to the vent stack (see *Figure 11*). **Circuit vents** and **loop vents** connect horizontal vents to the vent stack or stack vent. A circuit vent is a vent that connects a battery to the vent stack. In this arrangement, the circuit vent is connected to the drain before the first fixture and the last fixture and to the vent stack. The loop vent is a special type of circuit vent. Loop vents are installed from the fixture group to the stack vent. Some codes require that fixture batteries on the top floors of a building be vented with loop vents.

Select the proper type of battery vent for the application, and maintain the same pipe sizes throughout the vent. Local codes vary in their requirements for battery vents. Refer to your local code before venting a fixture battery.

4.4.0 Wet Vents

Vent piping that also carries liquid waste is a **wet vent**. It eliminates the need to install separate waste and vent stacks. Because of this, wet vents are a popular option for venting residential bathroom fixtures. Wet vents are installed either vertically or horizontally (see *Figure 12*). Only use wet vents with fixtures that have a comparatively low rate of flow, such as the following:

- Bathtubs
- Showers
- Bidets
- Lavatories
- Drinking fountains
- Water closets

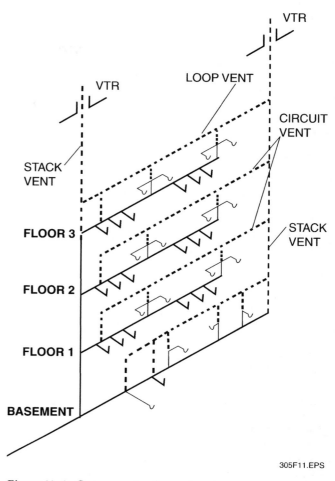

Figure 11 ◆ Components of a vent system.

 WARNING!
Be sure to follow your local code very carefully when installing wet vents. Ensure that the slope is correct. Otherwise, a mistake could cause serious drainage problems that could endanger public health.

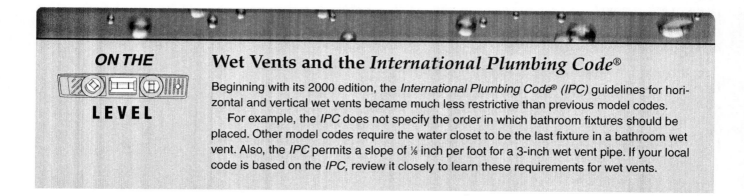

ON THE LEVEL

Wet Vents and the *International Plumbing Code*®

Beginning with its 2000 edition, the *International Plumbing Code*® (IPC) guidelines for horizontal and vertical wet vents became much less restrictive than previous model codes.

For example, the *IPC* does not specify the order in which bathroom fixtures should be placed. Other model codes require the water closet to be the last fixture in a bathroom wet vent. Also, the *IPC* permits a slope of ⅛ inch per foot for a 3-inch wet vent pipe. If your local code is based on the *IPC*, review it closely to learn these requirements for wet vents.

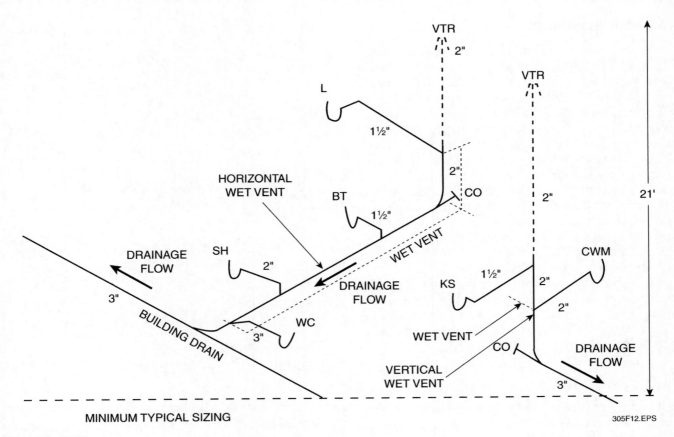

Figure 12 ◆ Wet vents.

When installing wet vents on toilets or water closets, follow the local code carefully. Make sure that the wet-vented fixture traps are capable of resealing. Maintain the proper hydraulic gradient and correct distances to ensure resealing.

Wet vents are considered acceptable by most model codes. Model codes cite the maximum vertical and horizontal run allowed for wet-vented pipe. The *Uniform Plumbing Code™ (UPC)* allows the use of wet vents for single and back-to-back bathroom fixtures. However, the *UPC* does not allow horizontal wet venting. Check your local code to determine the proper sizing and placement of a wet vent.

4.5.0 Air Admittance Vents

Air admittance vents ventilate the DWV stack using the building's own air (see *Figure 13*). Air admittance vents are also called air admittance valves. They are an alternative to vent stacks that penetrate building roofs. Air admittance vents operate like valves by allowing a controlled one-way flow of air into the drainage system. Reduced pressure causes the vent to open and allows air to enter the drain pipes. When the pressure is equalized, the vent closes by gravity. If back pressure occurs, the vent seals tightly to prevent sewer gases from escaping.

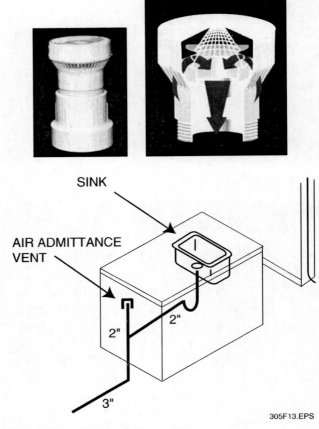

Figure 13 ◆ Air admittance vent.

5.12 PLUMBING ◆ LEVEL THREE

Air admittance vents come in two sizes. The large size vents entire systems. The smaller size vents individual fixtures. Local codes permit air admittance valves in residential buildings. The *IPC*, for example, allows air admittance valves on branch vents and individual vents. In the future, it may permit their use in stack vents as well. However, not all local codes accept air admittance valves. Check your local code for the proper sizing and placement of air admittance vents. Install air admittance vents as high as possible.

Use air admittance vents to vent fixtures in the center of a room, such as kitchen islands and wet bars. Codes cite maximum permitted distances between an island fixture and a main vent. Other options are to use a loop vent or a **combination waste and vent system** under the fixture (see *Figure 14*). This is a pipe that acts as a vent and also drains wastewater. If a combination waste and vent is used, ensure that the drain pipe is one size larger than the trap. The options for venting island fixtures depend on local codes.

4.6.0 Relief Vents

A **relief vent** allows excess air pressure in the drainage system to escape. This prevents back pressure from blowing out the seals in the fixture traps (see *Figure 15*). Relief vents also provide air to prevent trap loss due to pressure differential.

Use relief vents to increase circulation between the drain and vent systems. They can also serve as an auxiliary vent. A relief vent that is connected between the soil stack and the vent stack is called a **yoke vent.** Relief vents can also be connected between the soil stack and the fixture vents. In either case, connect the relief vent to the soil stack in a **branch interval,** which is the space between two branches entering a main stack. The space between branch intervals is usually story height but cannot be less than 8 feet.

Relief vents are very important in tall buildings. The weight of the water column in the stack can compress air trapped below it. Install relief vents on the floors of a high-rise building as called for in the design. A common practice is to install offset and yoke vents every 10 floors. Review your local code for specific details regarding the proper installation of relief vents.

In multistory buildings, a relief vent can often be combined with an offset in the waste stack to slow down the waste (see *Figure 16*). The installation will depend on the design of the stack. Local codes will specify the appropriate use of an offset relief vent in your area.

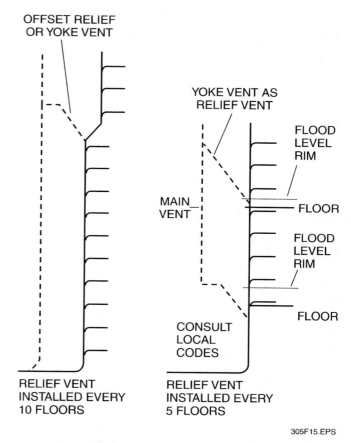

Figure 15 ♦ Relief vent.

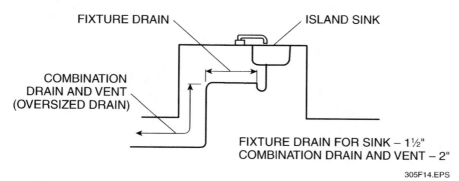

Figure 14 ♦ Combination waste and vent system.

4.7.0 Sovent® Vent Systems

An innovative vent design called the **Sovent® system** combines vent and waste stacks into one stack (see *Figure 17*). Where a branch joins the single stack, an aerator mixes waste from the branch with the air in the stack. This slows the velocity of the resulting mixture and keeps the stack from being plugged by wastewater. At the base of the stack, a de-aerator separates the air and the liquid. This relieves the pressure in the stack and allows the waste to drain out of the system. The Sovent® system is intended mostly for use in high-rise buildings.

Note that Sovent® systems are engineered vents. Some codes require approval by the project engineer before an engineered vent can be installed. Refer to your local code for the proper procedure in your area.

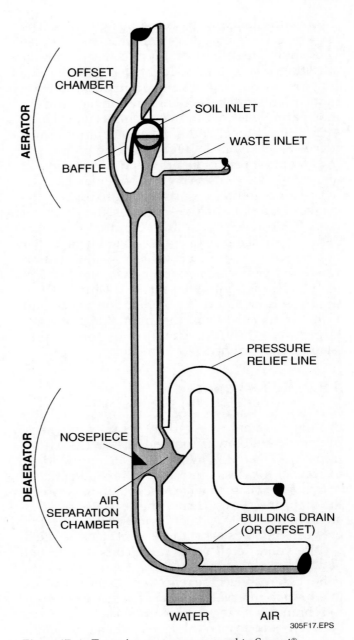

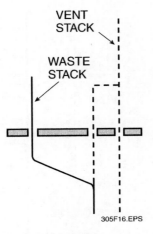

Figure 16 ◆ A relief vent at an offset.

Figure 17 ◆ Two-pipe system compared to Sovent® combination vent and waste system.

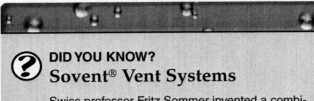

DID YOU KNOW?
Sovent® Vent Systems

Swiss professor Fritz Sommer invented a combination waste/vent system in 1959. Extensive tests proved the soundness of the basic design. Within a few years, Sovent® systems appeared in new buildings in Europe and the United States. Cast-iron, hubless fittings appeared in 1977. These cost much less than previous copper pipe and fittings. Advocates of Sovent® systems praise their simplicity, efficiency, and quiet operation.

ON THE LEVEL
Pneumatic Ejector Relief Vents

Do not connect a relief vent serving a pneumatic ejector to a fixture branch vent. Instead, carry the vent separately to a main vent or stack vent. Pneumatic ejector relief vents can also be vented to the open air.

Review Questions

Section 4.0.0

1. _____ vents run from a single fixture directly to the vent stack.
   a. Branch
   b. Individual
   c. Continuous
   d. Common

2. Continuous vents should be connected to the stack at least _____ inches above the fixture's flood level rim or overflow line.
   a. 4
   b. 6
   c. 8
   d. 12

3. Back vents may incorporate a continuous vent.
   a. True
   b. False

4. You are installing two fixtures back-to-back. Connect the fixtures to the common vent and drain with a short pattern sanitary _____.
   a. tee
   b. wye
   c. elbow
   d. cross

5. Fixture batteries can be vented using either a circuit vent or a _____ vent.
   a. loop
   b. relief
   c. wet
   d. line

6. A _____ vent connects a battery to the vent stack.
   a. common
   b. loop
   c. circuit
   d. battery

7. Wet vents are a popular option for venting _____.
   a. commercial water closets
   b. commercial lavatories
   c. island fixtures
   d. residential bathroom fixtures

8. A yoke vent is a type of _____ vent.
   a. battery
   b. individual
   c. relief
   d. island

9. To slow down flow in a waste stack, combine a relief vent with a(n) _____.
   a. aerator
   b. offset
   c. branch interval
   d. re-vent

10. The space between branch intervals is usually _____.
    a. story height but never less than 8 feet
    b. every other story height
    c. story height but never more than 8 feet
    d. every other story height but never less than 6 feet

Summary

Vents allow air to enter and exit a DWV stack to prevent pressure or a vacuum from disturbing fixture traps. Vents may ventilate single fixtures or a battery of them. They can provide relief venting to the main stack or connect with other vents. Regardless of how they are used, their purpose is to enable DWV systems to work by keeping an equal pressure inside the system. Vents come in many forms, and many can be used in combination with each other.

Like other plumbing installations, vents must be installed with proper grade and the correct pipe. Many codes differ as to how vents should be installed and where different types of vents should be used. Always consult your local code and the construction drawings prior to installing vents.

Consider this rule of thumb: *When in doubt, install a vent.*

Notes

Trade Terms Introduced in This Module

Air admittance vent: A valve-type vent that ventilates a stack using air inside a building. The vent opens when exposed to reduced pressure in the vent system, and it closes when the pressure is equalized.

Back pressure: Excess air pressure in vent piping that can blow trap seals out through fixtures. Back pressure can be caused by the weight of the water column or by downdrafts through the stack vent.

Back vent: An individual vent that is installed directly at the back of a single fixture and connects the fixture drain pipe to the vent stack or the stack vent.

Branch interval: The space between two branches connecting with a main stack. The space between branch intervals is usually story height but cannot be less than 8 feet.

Branch vent: A vent that connects fixtures to the vent stack.

Circuit vent: A vent that connects a battery of fixtures to the vent stack.

Combination waste and vent system: A line that serves as a vent and also carries wastewater. Sovent® systems are an example of a combination waste and vent system.

Common vent: A vent that is shared by the traps of two similar fixtures installed back-to-back at the same height. It is also called a unit vent.

Continuous vent: A vertical continuation of a drain that ventilates a fixture.

Flat vent: A vent that runs horizontally to a drain. It is also called a horizontal vent.

Hydraulic gradient: The maximum degree of allowable fall between a trap weir and the opening of a vent pipe. The total fall should not exceed one pipe diameter from weir to opening.

Indirect or momentum siphonage: The drawing of a water seal out of the trap and into the waste pipe by lower-than-normal pressure. It is the result of discharge into the drain by another fixture in the system.

Individual vent: A vent that connects a single fixture directly to the vent stack. Back vents and continuous vents are types of individual vents.

Loop vent: A vent that connects a battery of fixtures with the stack vent at a point above the waste stack.

Relief vent: A vent that increases circulation between the drain and vent stacks, thereby preventing back pressure by allowing excess air pressure to escape.

Re-vent: To install an individual vent in a fixture group.

Self-siphonage: A condition whereby lower-than-normal pressure in a drain pipe draws the water seal out of a fixture trap and into the drain.

Sovent® system: A combination waste and vent system used in high-rise buildings that eliminates the need for a separate vent stack. The system uses aerators on each floor to mix waste with air in the stack and a de-aerator at the base of the stack to separate the mixture.

Stack vent: An extension of the main soil and waste stack above the highest horizontal drain. It allows air to enter and exit the plumbing system through the building roof and is a terminal for other vent pipes.

Vent: A pipe in a DWV system that allows air to circulate in the drainage system, thereby maintaining equalized pressure throughout.

Vent stack: A stack that serves as a building's central vent, to which other vents may be connected. The vent stack may connect to the stack vent or have its own roof opening. It is also called the main vent.

Vent terminal: The point at which a vent terminates. For a fixture vent, it is the point where it connects to the vent stack or stack vent. For a stack vent, it is the point where the vent pipe ends above the roof.

Wet vent: A vent pipe that also carries liquid waste from fixtures with low flow rates. It is most frequently used in residential bathrooms.

Yoke vent: A relief vent that connects the soil stack with the vent stack.

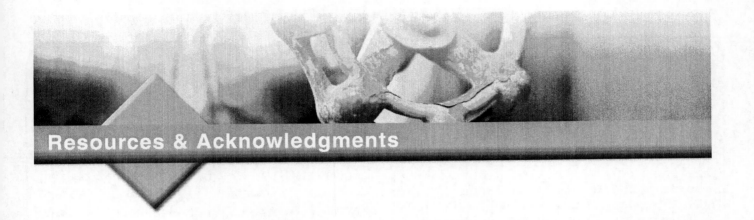

Resources & Acknowledgments

Additional Resources

This module is intended to be a thorough resource for task training. The following reference works are suggested for further study. These are optional materials for continued education rather than for task training.

Estimator's Man-Hour Manual on Heating, Air Conditioning, Ventilating, and Plumbing, 1978. John S. Page. Woburn, MA: Gulf Professional Publishing Company.

Planning Drain, Waste & Vent Systems, 1990. Howard C. Massey. Carlsbad, CA: Craftsman Book Company.

Plumbers and Pipefitters Handbook, 1996. William J. Hornung. Englewood Cliffs, NJ: Prentice Hall College Division.

References

Dictionary of Architecture and Construction, Third Edition. 2000. Cyril M. Harris, ed. New York: McGraw-Hill.

International Association of Plumbers and Mechanical Officials (IAPMO) website, www.iapmo.org, "2000 International Plumbing Code/Uniform Plumbing Code Review," Edward Saltzberg and J. Richard Wagner, http://www.iapmo.org/iapmo/code_comparison.html, viewed September 2004.

Pipefitters Handbook. 1967. Forrest R. Lindsey. New York: Industrial Press Inc.

Figure Credits

Copper Development Association, Inc., 305F16

International Code Council, Inc., 2000 IPC Commentary, 305F14

SE Sovent, 305F17

Studor, Inc., 305F12, 305F13A

CONTREN® LEARNING SERIES — USER UPDATE

The NCCER makes every effort to keep these textbooks up-to-date and free of technical errors. We appreciate your help in this process. If you have an idea for improving this textbook, or if you find an error, a typographical mistake, or an inaccuracy in NCCER's Contren® textbooks, please write us, using this form or a photocopy. Be sure to include the exact module number, page number, a detailed description, and the correction, if applicable. Your input will be brought to the attention of the Technical Review Committee. Thank you for your assistance.

Instructors – If you found that additional materials were necessary in order to teach this module effectively, please let us know so that we may include them in the Equipment/Materials list in the Annotated Instructor's Guide.

Write: Product Development and Revision
National Center for Construction Education and Research
P.O. Box 141104, Gainesville, FL 32614-1104

Fax: 352-334-0932

E-mail: curriculum@nccer.org

Craft _____ Module Name _____

Copyright Date _____ Module Number _____ Page Number(s) _____

Description _____

_____

_____

_____

_____

(Optional) Correction _____

_____

_____

_____

(Optional) Your Name and Address _____

_____

_____

_____

Plumbing Level Three

02306-06

Sizing DWV and Storm Systems

02306-06
Sizing DWV and Storm Systems

Topics to be presented in this module include:

1.0.0 Introduction 6.2
2.0.0 Sizing Drain, Waste, and Vent Systems 6.2
3.0.0 Sizing Storm Drainage Systems 6.13

Overview

Safe and efficient drainage of wastes is one of the most important functions of plumbing systems. Drain, waste, and vent (DWV) systems and storm sewers remove these wastes. Codes require DWV and storm drainage systems in all commercial and residential buildings; specific requirements vary from region to region. Plumbers are responsible for sizing and installing both types of systems, and must custom design each one.

To size a DWV system, plumbers determine the discharge rates in drainage fixture units (DFUs) for all drains, stacks, sewer lines, and vents in the system. DFUs measure the waste discharge of fixtures; the more DFUs a fixture discharges, the larger the fixture drain pipe needs to be. Local plumbing codes provide many tables and guidelines to assist in sizing system components. Based on DFU figures, plumbers then determine pipe sizes for vertical stacks and horizontal fixture branches. To size building drains and sewers, the DFU measurement is also used along with the pipe slope. When sizing vents, plumbers consider the number and types of fixtures in the system, the number of DFUs being sent to the drainage system, and the length of vent pipes.

Storm drainage systems are required for building roofs, paved areas, and residential lawns and yards. Correct sizing of storm systems depends upon the maximum expected rainfall during a given time period and the designed rate of removal for the structure. Because public health is at stake, plumbers must be very careful to size each system correctly.

Focus Statement
The goal of the plumber is to protect the health, safety, and comfort of the nation job by job.

Code Note
Codes vary among jurisdictions. Because of the variations in code, consult the applicable code whenever regulations are in question. Referring to an incorrect set of codes can cause as much trouble as failing to reference codes altogether. Obtain, review, and familiarize yourself with your local adopted code.

Portions of this publication reproduce tables and figures from the *2003 International Plumbing Code* and the *2000 IPC Commentary*, International Code Council, Inc., Falls Church, Virginia. Reproduced with permission. All rights reserved.

Objectives

When you have completed this module, you will be able to do the following:

1. Calculate drainage fixture units for waste systems.
2. Size building drains and sewers.
3. Size a vent system.
4. Identify and size special kinds of waste and vent systems.
5. Size roof drainage systems.

Trade Terms

Conductor
Controlled-flow roof drainage system
Conventional roof drainage system
Drainage fixture unit (DFU)
Percolate
Permeability
Ponding
Scupper
Secondary drain

Required Trainee Materials

1. Appropriate personal protective equipment
2. Sharpened pencils and paper
3. Copy of local applicable code
4. Calculator

Prerequisites

Before you begin this module, it is recommended that you successfully complete *Core Curriculum; Plumbing Level One; Plumbing Level Two; Plumbing Level Three*, Modules 02301-06 through 02305-06.

This course map shows all of the modules in the third level of the *Plumbing* curriculum. The suggested training order begins at the bottom and proceeds up. Skill levels increase as you advance on the course map. The local Training Program Sponsor may adjust the training order.

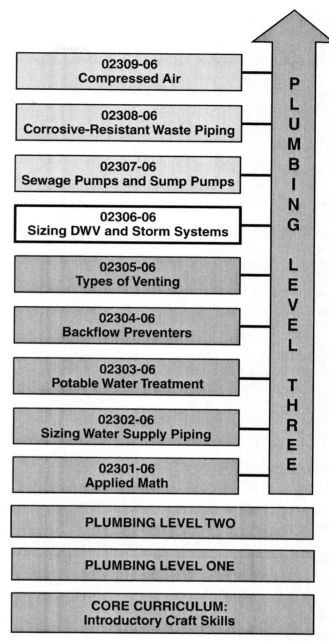

1.0.0 ◆ INTRODUCTION

Safe and efficient drainage is one of the most important functions of plumbing systems. Public health depends on the removal of sewage and storm wastes. Drain, waste, and vent (DWV) systems and storm sewers remove these wastes. Plumbers are responsible for sizing and installing both types of systems. You have already learned how to install different types of pipe. In this module, you will learn to apply your math and code-reading skills to select the proper size of pipe for sewage and storm water drainage.

Codes require DWV and storm drainage systems in all commercial and residential buildings. As with other types of plumbing installations, each system is unique. Plumbers custom design systems for each installation. Code requirements for DWV and storm drainage vary from region to region. Take the time to study your local code.

Note that the tables shown in this module are specific examples from real model codes. They are provided for reference only. Do not use the tables in this module to size actual DWV and storm drainage systems in the field! Always refer to the appropriate information in your local applicable code.

2.0.0 ◆ SIZING DRAIN, WASTE, AND VENT SYSTEMS

Earlier you learned how to install and test DWV systems. In this section, you will learn how to size the pipe used in those systems. Sizing drains, stacks, sewer lines, and vents is generally a paper and pencil task. You have to sit down and calculate the proper pipe sizes before you can begin to build the system. You do not have to be an engineer to be able to size a DWV system so that it works. Local plumbing codes provide many tables and guidelines that allow you to size the system components. You will learn about those guidelines and practice using them as part of this module. Terms and definitions vary from jurisdiction to jurisdiction. Be sure to consult your local code for the proper terms in your area.

You have already learned that vents work in conjunction with drains to enable wastes to flow from fixtures into the waste pipe, where they are carried into the sewer. Without vents, plumbing systems would not work. Every structure with a plumbing system requires at least one main vent. Regardless of the type of installation, consider the following factors when sizing vents:

- The number and types of fixtures attached to the system
- The number of **drainage fixture units** (DFUs) being transferred to the drainage system
- The length of the vent pipe

2.1.0 Calculating Drainage Fixture Units

Pipe size in a DWV system is based on the number of drainage fixture units the system carries. The larger the number of DFUs, the greater the pipe diameter required. DFUs measure the waste discharge of fixtures (see *Table 1*). DFUs are determined based on the following variables:

- The fixture's rate of discharge
- The duration of a single discharge
- The average time between discharges

When a DFU value has not been determined, use the standard value of 7.5 gallons of discharge per minute.

Always consult your local applicable codes to find the values in use in your area.

The more DFUs a fixture discharges, the larger the fixture drain pipe needs to be. The fixtures listed in *Table 1* pass a variable quantity of water depending on how frequently they are used. For fixtures that drain waste continuously, such as air conditioners, codes will assign a minimum value for each gpm of discharge.

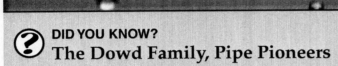

> **DID YOU KNOW?**
> **The Dowd Family, Pipe Pioneers**
>
> At the dawn of the 20th century, W. Frank Dowd opened a foundry in Charlotte, North Carolina, to produce cast-iron soil pipe. Originally, Charlotte Pipe employed 25 men. Dowd led the company until 1926, when his son Frank Dowd II took over and expanded the business. In the 1950s, Frank II's sons Frank Jr. and Roddey took over. They explored new production technologies and experimented with new pipe materials, including plastics. Today, Charlotte Pipe is the country's largest producer of DWV pipe and fittings. It also holds the record for the longest period of any pipe manufacturer under single ownership—the Dowd family, innovators in the pipe manufacturing field.

Table 1 DFUs for Fixtures and Groups

Fixture Type	DFU Value as Load Factors	Minimum Diameter of Trap (in.)
Automatic clothes washers, commercial[a,g]	3	2
Automatic clothes washers, residential[g]	2	2
Bathroom group (water closet, lavatory, bathtub or shower, with or without a bidet, and an emergency drain), 1.6 gal per flush water closet[f]	5	—
Bathroom group (as above), >1.6 gal per flush water closet[f]	6	—
Bathtub[b] (with or without overhead shower or whirlpool attachments)	2	1½
Bidet	1	1¼
Combination sink and tray	2	1½
Dental lavatory	1	1¼
Dental unit or cuspidor	1	1¼
Dishwashing machine[c], domestic	2	1½
Drinking fountain	½	1¼
Emergency floor drain	0	2
Floor drains	2	2
Kitchen sink, domestic	2	1½
Kitchen sink, domestic, with food waste grinder and/or dishwasher	2	1½
Laundry tray (1 or 2 compartments)	2	1½
Lavatory	1	1¼
Shower	2	1½
Sink	2	1½
Urinal	4	(note d)
Urinal, ≤1gal per flush	2[e]	(note d)
Wash sink (circular or multiple), each set of faucets	2	1½
Water closet, flushometer tank, public or private	4[e]	(note d)
Water closet, private (1.6 gal per flush)	3[e]	(note d)
Water closet, private (>1.6 gal per flush)	4[e]	(note d)
Water closet, public (1.6 gal per flush)	4[e]	(note d)
Water closet, public (>1.6 gal per flush)	6[e]	(note d)

[a] This is not applicable for traps larger than 3 inches.
[b] A shower head over a bathtub or whirlpool bathtub attachment does not increase the DFU value.
[c] See local code for methods of computing unit value of fixtures not listed or for rating of devices with intermittent flows.
[d] Trap size shall be consistent with fixture outlet size.
[e] For the purpose of computing loads on building drains and sewers, water closets and urinals shall not be rated at a lower DFU unless the lower values are confirmed by testing.
[f] For fixtures added to a dwelling unit bathroom group, add the DFU value of those additional fixtures to the bathroom group fixture count.
[g] See your local code for sizing requirements for fixture drain, branch drain, and drainage stack for an automatic clothes washer standpipe.

Source: International Code Council, Inc.

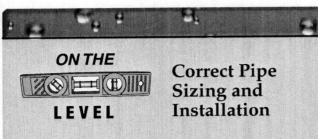

ON THE LEVEL

Correct Pipe Sizing and Installation

You have learned that undersizing a pipe can be a costly mistake. However, oversizing can be just as costly. If an installer uses a larger pipe size than necessary, water may not flow fast enough in the horizontal section to scour the pipe and wash away wastes. When installed at an incorrect slope, an oversized pipe can clog even faster than an undersized one. Take the time to review your local code requirements for pipe sizing. Calculate pipe sizes using the most recent standards. The results will help your customers.

2.2.0 Sizing Drains

After determining the DFUs for all the fixtures in the structure, you can begin to size the drainage system. Your local code will provide tables that show the correct pipe size based on the number of DFUs in the system (see *Table 2*). Use this information to determine pipe sizes for both vertical stacks and horizontal fixture branches.

For example, *Table 1* assigns 2 DFUs to a bathtub with or without attachments. Assume that the bathtub is the only fixture on the horizontal branch. Use *Table 2* to calculate the correct pipe size. In this case, the pipe should be 1½ inches (because the DFU rating of 2 is not listed, size up to the next listed DFU rating). *Table 1* also requires that the minimum trap size for a bathtub be 1½ inches. Note that you can also use fixture group sizing methods for bathroom groups. These values are based on a probability of usage. This is because the fixtures are not likely to be used all at the same time.

The number of water closets on a line and the number of floors in the structure can also affect drainage pipe sizing. If a single bathroom group has more than one water closet, you must increase the size of the drain stack. That way, the stack can handle the extra surge if both water closets are flushed at the same time. Follow your local code closely when installing stacks in buildings that require three or more branch intervals. Remember also to consider future fixture installations when sizing pipes. Install branch stubs for these. Cap the stubs with approved caps or plugs.

Table 2 Sample Table for Calculating the Size of Horizontal Branches and Vertical Stacks

Diameter of Pipe (in.)	Total for Horizontal Branch	Maximum Number of DFUs — Stacks[b]		
		Total Discharge into One Branch Interval	Total for Stack of Three Branch Intervals or Less	Total for Stack Greater Than Three Branch Intervals
1½	3	2	4	8
2	6	6	10	24
2½	12	9	20	42
3	20	20	48	72
4	160	90	240	500
5	360	200	540	1,100
6	620	350	960	1,900
8	1,400	600	2,200	3,600
10	2,500	1,000	3,800	5,600
12	2,900	1,500	6,000	8,400
15	7,000	(note c)	(note c)	(note c)

[a] Does not include branches of the building drain.
[b] Stacks shall be sized on the total accumulated connected load at each story or branch interval. As the total accumulated connected load decreases, stacks are permitted to be reduced in size. Stack diameters shall not be reduced to less than one-half of the diameter of the largest stack size required.
[c] Sizing load based on design criteria.

Source: International Code Council, Inc.

2.2.1 Exceptions and Limitations

The guidelines given in the previous sections have many exceptions. In every case, consult your local code. Discuss tips and techniques with experienced plumbers. Learn about the limitations and exceptions as well as the general rules. Because the health of the public is at stake, take the time to size each system correctly.

Many codes require drainage pipes that run underground or below a basement or cellar to be at least 2 inches in diameter. These codes require this regardless of the DFU values that have been calculated for the system. Horizontal drains should be sloped to allow the wastewater to flow at least 2 feet per second. If a drain line is less than 4 inches in diameter, the slope must be ¼ inch per foot of pipe. If a drain line is larger than 4 inches in diameter, the slope must be ⅛ inch per foot. Water closets require a minimum drain size of 3 inches. Some codes require 4-inch pipe if three or more water closets are installed on a single drain line.

Use at least a 3-inch-diameter pipe when installing soil stacks. Codes prohibit using branch piping that has a larger diameter than the soil stack it connects to. Many codes also limit how many branches can be connected to a 3-inch stack. Usually, no more than two branches per floor or six branches per stack are permitted.

Often, the location of a building's structural components makes it impossible to install drainage stacks that run straight from top to bottom. Install offsets to route the stacks around obstacles. In most codes, if the offset is 45 degrees or less, the pipe can be the same size as the vertical stack. If the offset is greater than 45 degrees, size the pipe according to the rules for sizing horizontal pipe. If the offset is installed above the highest fixture branch, size the offset according to the rules for sizing vents. Before connecting horizontal branch lines above and below offsets, consult your local code.

2.2.2 Sizing Building Drains and Sewers

Like horizontal branches and vertical stacks, building drains and sewers are sized according to the DFUs assigned to them (see *Table 3*). Additionally, consider the pipe slope when sizing building drains and sewers. The slope may range from 1/16 to ½ inch per foot. The combination of pipe diameter and slope determines the carrying capacity of every building drain and sewer.

Local codes limit the number of branches and water closets on a building drain or sewer. Most codes require that building sewer lines be at least 4 inches in diameter. Some codes allow the slope of the building drain or sewer to be increased so that it can handle a greater number of DFUs. For example, referring to *Table 2*, an 8-inch pipe can handle 1,400 DFUs at a slope of 1/16 inch per foot. If the slope is increased to ½ inch per foot, however, the same size pipe can handle 2,300 DFUs. Refer to your local code for the standards that apply in your area.

Table 3 Sample Table for Calculating the Size of Building Drains and Sewers

Diaeter of Pipe (in.)	Maximum Number of DFUs Connected to Any Portion of the Building Drain or Building Sewer, Including Branches of the Building Drain[a]			
	Slope per Foot			
	1/16 inch	⅛ inch	¼ inch	½ inch
1¼	—	—	1	1
1½	—	—	3	3
2	—	—	21	26
2½	—	—	24	31
3	—	36	42	50
4	—	180	216	250
5	—	390	480	575
6	—	700	840	1,000
8	1,400	1,600	1,920	2,300
10	2,500	2,900	3,500	4,200
12	3,900	4,600	5,600	6,700
15	7,000	8,300	10,000	12,000

[a] The minimum size of any building drain serving a water closet shall be 3 inches.

Source: International Code Council, Inc.

ON THE LEVEL

The Mechanics of Fluid Flow

In horizontal branch lines, water flows along the lower portion of the pipe. Air circulates in the upper portion. You have already learned that horizontal pipes are sloped so that water can flow through the pipe at a rate of at least 2 feet per second. This rate of flow sets up a scouring action in the pipe and prevents sediment from building up in the pipe.

The mechanics of vertical flow are different from those of horizontal flow. When water enters the vertical stack and flows down to the building drain, it runs along the wall of the pipe. The velocity of water in a vertical stack is usually about 10 to 15 feet per second.

As the flow in the stack increases, the space for air decreases. When the stack is about one-third full of water, it reaches its maximum flow capacity. Above this level, slugs of water begin to drop away from the pipe wall and fall through the middle of the stack. The slugs eventually land back on the pipe wall further down. This is called diaphragming. Diaphragming causes many of the noises heard in soil stacks.

2.3.0 Sizing Vents

Earlier you learned about the different types of vents and how they work. In this section, you will learn the general techniques for sizing vents. Always refer to your local code for specific requirements in your area. Always ensure that all joints in vent piping are sealed correctly. Leaks allow odors to escape, and escaping sewer gases can cause explosions, contamination and illness.

In cold climates and regions with heavy snow, vent terminals can become plugged by frost or covered by a snow blanket. This is caused by water in the vent air condensing and building up on the inside of the vent pipe as it comes into contact with the cold outside air. To prevent frost buildup in the vent pipe, enlarge the section of pipe that extends above the roof (see *Figure 1*). A pipe at least 3 inches in diameter is less likely to completely close up with frost than a narrower pipe. The enlargement should extend at least 1 foot inside the roof as well. Ensure that the location of the vent and the vent pipe materials are suitable for the climate conditions. Otherwise, the same problem may happen in the future.

Calculate diameter for a vent by measuring the developed length of the vent and the total number of DFUs connected to it. Your local code should have a table that allows you to calculate this (see *Table 4*).

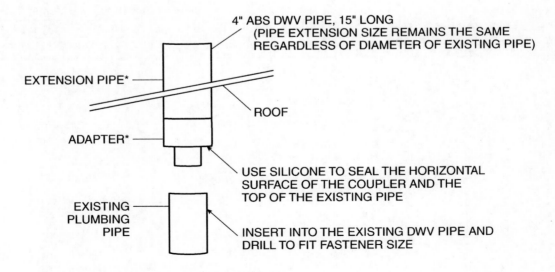

Figure 1 ◆ Enlarged vent extension designed to prevent frost closure.

Table 4 Size of Individual Vent Determined by Total Length and Drainage Fixture Units

Diameter of Soil or Waste Stack (in.)	Total DFUs Being Vented	Maximum Developed Length of Vent (ft.)[a] Diameter of Vent (in.)										
		1¼	1½	2	2½	3	4	5	6	8	10	12
1¼	2	30										
1½	8	50	150									
1½	10	30	100									
2	12	30	75	200								
2	20	26	50	150								
2½	42		30	100	300							
3	10		42	150	360	1,040						
3	21		32	110	270	810						
3	53		27	94	230	680						
3	102		25	86	210	620						
4	43			35	85	250	980					
4	140			27	65	200	750					
4	320			23	55	170	640					
4	540			21	50	150	580					
5	190				28	82	320	990				
5	490				21	63	250	760				
5	940				18	53	210	660				
5	1,400				16	49	190	590				
6	500					33	130	400	1,000			
6	1,100					26	100	310	780			
6	2,000					22	84	260	660			
6	2,900					20	77	240	600			
8	1,800						31	95	240	940		
8	3,400						24	73	190	720		
8	5,600						20	62	160	610		
8	7,600						18	56	140	560		
10	4,000							31	78	310	960	
10	7,200							24	60	240	740	
10	11,000							20	51	200	630	
10	15,000							18	46	180	570	
12	7,300								31	120	380	940
12	13,000								24	94	300	720
12	20,000								20	79	250	610
12	26,000								18	72	230	500
15	15,000									40	130	310
15	25,000									31	96	240
15	38,000									26	81	200
15	50,000									24	74	180

[a] The developed length shall be measured from the vent connection to the open air.

Source: International Code Council, Inc.

In most cases, the diameter of a vent should be one-half the diameter of the drain being vented. However, the vent diameter should never be less than 1¼ inches. If the vent is longer than 40 feet, use one pipe size larger than the calculated size for the entire length. Measure the developed length from the farthest drainage connection to the stack or terminal connection. Review your local code to find out how to calculate vent sizes in your area.

2.3.1 Main Vents

The main vent is the vent stack that carries the largest number of DFUs in the system. Usually, the location of the water closet determines the location of the main vent. Most installations require a main vent that is at least 3 inches in diameter. Ensure that the main vent meets local code requirements. If the DFUs exceed the code specifications, install a larger vent.

If a vent for a washing machine or laundry tray is installed in an unattached garage, use 1½-inch pipe for the main vent. Buildings that have many separate units, such as shopping malls and condominiums, may also require smaller vents. Install a main vent for each individual unit. Consult your local code for sizing requirements.

Use the same diameter pipe for the entire length of the main vent. Measure the developed length from the point where the vent intersects the stack or drain to the point where it terminates through the roof. Note that in some designs, the main vent may intersect the main stack before it exits the structure. In such cases, be sure to include the length of all sections that are required for the vent to reach the open air (see *Figure 2*).

2.3.2 Individual Vents

Individual vents connect single fixtures to the main vent (see *Figure 3*). They are very dependable and can be used on remote fixtures where re-venting is not practical. The size of an individual vent depends on the size and DFU rating of the fixture to be attached. Individual vents must be at least half the size of the fixture drain and are never less than 1¼ inches in diameter. Consult your local code for other requirements. Individual vents are often called back vents or continuous vents.

2.3.3 Common Vents

Common vents are used to vent two fixtures that are set back to back or side by side on the same floor (see *Figure 4*). The fixture vents can be connected to the common vent at the same level or at different levels. If the vents are connected at the

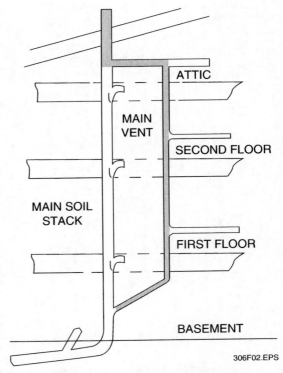

Figure 2 ◆ Measuring the developed length of a vent.

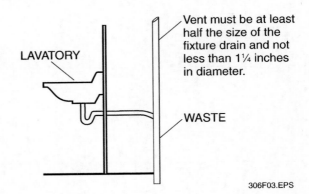

Figure 3 ◆ Individual vent.

same level (see *Figure 4A*), locate the vents at the intersection of the fixture drains or downstream of them. If the vents are connected at different levels, the upper fixture drain must be one pipe diameter larger than the vertical drain. The lower fixture drain, however, should not be larger than the vertical drain (see *Figure 4B*). As a rule of thumb, do not allow more than 1 DFU on a 1½-inch pipe or more than 4 DFUs on a 2-inch pipe when sizing common vents.

2.3.4 Wet Vents

To determine the correct pipe size for a wet vent, add the DFUs for all the fixtures in the fixture group. Refer to *Table 1* or the appropriate table in your local code to determine the DFU load. Then

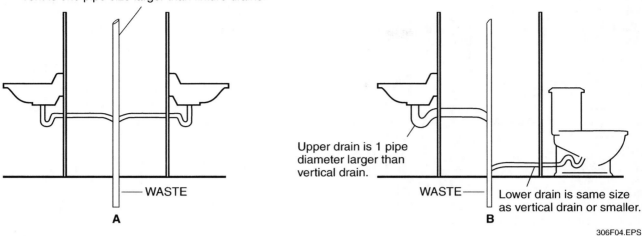

Figure 4 ♦ Typical common vent installation.

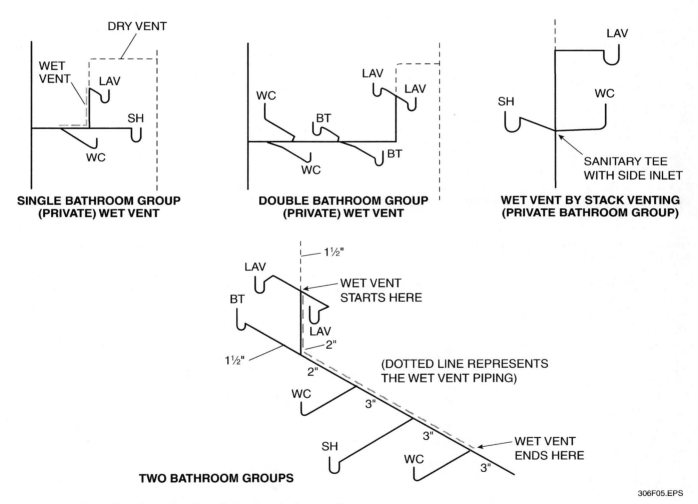

Figure 5 ♦ Examples of wet vent installation in a bathroom fixture group.

refer to *Table 5* or your local code to select the proper vent size. Do not install wet vents on wall-hung water closets.

For example, if a lavatory drain is being used to vent a bathtub and water closet (see *Figure 5*), the lavatory drain must be larger than the normal required size. Calculate the total DFUs for all three fixtures. According to *Table 1*, a total of 6 DFUs should be assigned to the group. According to *Table 5*, a 2½-inch pipe will be required to vent the group.

Table 5 Sample Table for Sizing Wet Vents

Wet Vent Pipe Size (in.)	Load (DFUs)
1½	1
2	4
2½	6
3	12

Source: International Code Council, Inc.

2.3.5 Relief Vents

Relief vents provide adequate airflow in cases where the number of fixtures installed becomes too large for conventional venting methods. Install relief vents, also called yoke vents, in all soil and waste stacks that have more than 10 branch intervals. Install relief vents at every 10th branch, counting from the top floor down. The relief vent must be the same size as the vent stack to which it connects.

2.3.6 Battery Vents

Use a circuit vent to vent a battery of fixtures (see *Figure 6*). Consult your local code for sizing information. Circuit vents are sized differently than other types of vents. A circuit vent should be at least one-half the size of the horizontal waste line that it is venting, but it must never be less than 1½ inches in diameter. Do not install more than eight

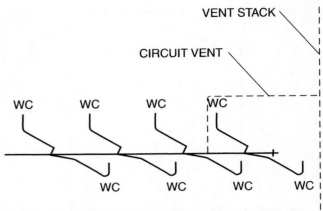

BACK-TO-BACK WATER CLOSETS CIRCUIT VENTED

Figure 6 ◆ Typical installation of circuit vents.

floor outlet water closets, floor outlet urinals, showers, bathtubs, or floor drains to a single circuit vent. However, if the water closets are blowout types, more than eight can be installed. If lavatories are installed above a circuit vent, install a continuous vent on each vertical branch.

2.3.7 Sump Vents

The venting of drain pipes that are below the sewer level is similar to the venting of gravity-fed systems. Consult your local code for the correct sizes and lengths (see *Table 6*). For pneumatic ejec-

Table 6 Sample Sizing Table for Sump Vents

Discharge Capacity of Pump (gpm)	Maximum Developed Length of Vent (ft.)[a] Diameter of Vent (in.)					
	1¼	1½	2	2½	3	4
10	No limit (note b)	No limit	No limit	No limit	No limit	No limit
20	270	No limit	No limit	No limit	No limit	No limit
40	72	160	No limit	No limit	No limit	No limit
60	31	75	270	No limit	No limit	No limit
80	16	41	150	380	No limit	No limit
100	10[c]	25	97	250	No limit	No limit
150	Not permitted	10[c]	44	110	370	No limit
200	Not permitted	Not permitted	20	60	210	No limit
250	Not permitted	Not permitted	10	36	132	No limit
300	Not permitted	Not permitted	10[c]	22	88	380
400	Not permitted	Not permitted	Not permitted	10[c]	44	210
500	Not permitted	Not permitted	Not permitted	Not permitted	24	130

[a] Developed length plus an appropriate allowance for entrance losses and friction due to fittings, changes in direction and diameter. Suggested allowances shall be obtained from NSB Monograph 31 or other approved sources. An allowance of 50 percent of the developed length shall be assumed if a more precise value is not available.
[b] Actual values greater than 500 feet.
[c] Less than 10 feet.

Source: International Code Council, Inc.

tors, the vent should allow the ejector to reach normal air pressure. Connect vents for pneumatic ejectors to an independent vent stack that terminates through the roof. Do not size the vent smaller than 1¼ inches in diameter. Connect vents for pneumatic sewage ejectors to an independent vent stack. This vent stack must be at least 1¼ inches in diameter.

2.3.8 Combination Drain and Vent Systems

Use combination drain and vent systems for floor drains, standpipes, sinks, and lavatories (see *Figure 7*). Do not vent urinals or water closets using a combination drain and vent. Kitchen sinks and washing machine traps may not be installed on a 2-inch combination drain and vent. Consult your local code for sizing tables for combination drain and vent installations (see *Table 7*).

Table 7 Sample Sizing Table for Combination Waste and Vent Systems

Pipe Diameter (in.)	Maximum Number of DFUs	
	Connecting to a Horizontal Branch or Stack	Connecting to a Building Drain or Subdrain
2	3	4
2½	6	26
3	12	31
4	20	50
5	160	250
6	360	575

Source: International Code Council, Inc.

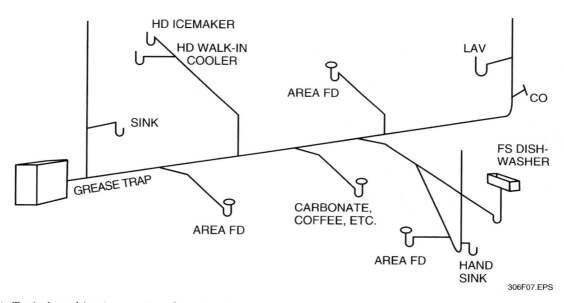

Figure 7 ◆ Typical combination waste and vent system.

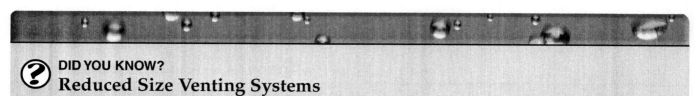

? DID YOU KNOW?
Reduced Size Venting Systems

Some codes allow the use of a reduced size venting system. This is an engineered system that is used when every fixture is individually vented and connected to either a branch vent or a vent stack. In reduced size venting systems, the vent pipe size can be as small as ½ inch. This applies only to the length of each individual pipe. The airflow rate for the reduced size vent should be calculated before installation.

The local code provides information for sizing these vents. In general, if a construction project involves a reduced size venting system, the engineer will give instructions for how it is to be installed.

ON THE LEVEL

The National Pollutant Discharge Elimination System (NPDES) Program

The United States Environmental Protection Agency (EPA) controls water pollution through its NPDES program. NPDES regulates point sources, such as pipes and ditches, that discharge pollutants into surface water. Industrial, municipal, and other facilities must obtain permits from the EPA if they discharge directly into surface water. In most cases, the NPDES permit program is administered at the state level.

The NPDES permit program has helped to improve the country's water quality significantly. Twenty-five years ago, only about one-third of our surface waters were safe for fishing and swimming. Today, the percentage has risen to two-thirds. Loss of wetlands has been reduced by more than 80 percent. Soil erosion from agricultural runoff has been nearly cut in half.

Review Questions

Sections 1.0.0–2.0.0

1. DFUs measure the _____ of fixtures.
   a. cubic footage
   b. fresh water requirement
   c. waste discharge
   d. continuous flow

2. If a drain line is less than 4 inches in diameter, the slope must be _____ inch per foot of pipe.
   a. ½
   b. ¼
   c. ⅛
   d. ¹⁄₁₆

3. What is the maximum developed length of vent on a 3-inch waste stack with a fixture load of 40 DFUs and a 2-inch vent? (Refer to Table 4.)
   a. 150
   b. 300
   c. 27
   d. 94

4. Measure the developed length of a main vent from the point where the vent intersects the stack or drain to the point where the vent _____.
   a. intersects the main stack
   b. intersects the highest fixture branch
   c. terminates in a circuit or loop vent
   d. terminates through the roof

5. Do not install more than _____ floor outlet urinals to a single circuit vent.
   a. eight
   b. six
   c. four
   d. two

6. If two fixture vents are connected to a common vent at different levels, the upper fixture drain must be _____ the vertical drain.
   a. the same diameter as
   b. one DFU smaller than
   c. one pipe diameter larger than
   d. ½ pipe diameter larger than

3.0.0 ◆ SIZING STORM DRAINAGE SYSTEMS

Correct sizing of storm drainage systems depends upon two factors. The first is the maximum expected rainfall during a given period of time. The second is the designed rate of removal. Local codes provide guidance on how to size and install storm water drainage systems. Storm drainage systems are required for the following:

- Building roofs
- Paved areas, such as parking lots and streets
- Residential lawns and yards

For one- and two-family homes, channel the storm water away from the house and onto lawns or roads. Many codes require cleanouts on storm drainage systems.

Storm water can be disposed of right away, or it can be stored and then released slowly. Consult your local code to learn how storm water disposal is handled in your area. Designers can choose one of three methods for retaining and disposing of storm water. Select the method that is appropriate for the type of structure and the anticipated amount of drainage:

- Above-grade drainage
- Below-grade drainage
- Roof drainage

Depending on your local code, connect storm drains to a combined sewer line or to a separate storm sewer. Never connect storm drainage to a sewage-only line. If the storm drain connects to a combined sewer line, install a trap. Traps can be installed in two ways. One option is to trap all drain branches that serve each **conductor**. A conductor is a spout or vertical pipe that drains storm water into the main drain line; conductors are also called leaders. The other option is to install a trap in the main drain immediately upstream of its connection with the sewer. Do not reduce the size of a drainage pipe in the direction of flow.

 WARNING!
Most codes will not allow storm drainage to be connected to a sewage-only line, because the occasional heavy volume in storm drains could overwhelm a sewage system. In addition, storm drainage contains numerous dangerous materials.

Building engineers usually design storm drainage systems. They must ensure that the building is strong enough to handle the weight of water. Follow the design closely when installing the system so that you will know that it works efficiently and effectively. Always refer to your local code.

3.1.0 Rainfall Conversion

Consider the annual amount of rainfall in your region when sizing storm drains. Also consider the potential for flash floods. Unlike most other estimates that you will make, rainfall calculations are based on probabilities, not facts. Nevertheless, rainfall probabilities are a reliable tool for sizing storm drainage systems. Your local code will have copies of rainfall maps for your area. Most model and local codes get their climate information from the National Oceanic and Atmospheric Administration (NOAA). You can find NOAA online at www.noaa.gov.

The local 30-year average is the most accurate indicator of rainfall. It takes both wet and dry years into consideration. The system must also be designed to handle the most severe storm that is likely to occur. Estimates for this kind of storm are calculated for 10-, 25-, 50-, and 100-year periods (see *Figure 8*). The most severe rainstorm would likely occur only once every 100 years (see *Table 8*). Of course, this does not tell you when during that 100-year period the storm will happen. It may happen in the first year, it may happen in the hundredth year, or it may not happen at all. However, the system should be able to handle even this rare storm.

To size a drainage system, find the area of the surface and the volume of water it can hold. For example, you have to drain a roof that is 50 feet wide by 100 feet long. You know that its total area is 5,000 square feet (50 feet × 100 feet). Now imagine that the design calls for the roof to hold 6 inches of water. The roof will hold up to 2,500 cubic feet of water (50 feet × 100 feet × 0.5 foot). Converted to gallons, that equals 18,750 gallons (2,500 cubic feet × 7.5 gallons per cubic foot).

The building engineer has to make sure that a roof or other storage area can hold the weight of water. In the example above, 18,750 gallons of water weigh more than 156,000 pounds (18,750 gallons × 8.33 pounds per gallon). That is more than 78 tons of water. You can see why it is important to size a retention and drainage system correctly.

Just 2 inches of rainfall in an hour weighs more than 10 pounds (see *Table 9*). The same amount of rainfall equals 1.25 gallons (see *Table 10*). Use *Tables 9* and *10* to calculate the weight and amounts of rain that a drainage system will have to handle.

Figure 8 ◆ Sample map showing 100-year, 1-hour rainfall in inches.

Table 8 Sample Table of Precipitation Intensity Estimates, in Inches per Hour

Location	10-Year Storm		25-Year Storm		50-Year Storm		100-Year Storm	
	1 hour	24 hours	1 hour	24 hours	1 hour	24 hours	1 hour	24 hours
Columbia, South Carolina	2.60	0.22	3.06	0.26	3.42	0.30	3.79	0.34
Moorestown, New Jersey	2.08	0.21	2.47	0.25	2.76	0.29	3.06	0.34
Chillicothe, Illinois	2.16	0.18	2.56	0.21	2.86	0.24	3.16	0.27
Santa Fe, New Mexico	1.26	0.09	1.54	0.11	1.74	0.12	1.97	0.13

Table 9 Weight of Rainwater

Rainfall (in./hr.)	Weight (lb./sq. ft.)
1	5.202
2	10.404
3	15.606
4	20.808
5	26.010
6	31.212

Table 10 Rainfall Conversion Data for Calculating Water Volume

Rainfall (in./hr.)	Gallons per Minute (per sq. ft.)	Gallons per Hour (per sq. ft.)
1	0.0104	0.623
2	0.0208	1.247
3	0.0312	1.870
4	0.0416	2.493
5	0.0520	3.117
6	0.0624	3.740

3.2.0 Sizing Roof Storage and Drainage Systems

As mentioned, storm water can be drained immediately, or it can be stored and released over a longer period. Systems that drain immediately are called **conventional roof drainage systems**. Systems that store water and release it slowly are called **controlled-flow roof drainage systems**. In this section, you will learn how to size both types of drainage systems.

3.2.1 Sizing Conventional Roof Drainage Systems

The conventional way to drain storm water from a roof is to allow the water to drain as soon as it falls. This type of drainage system must be able to handle the largest expected rainfall in the area. Drainage systems consist of vertical leaders and horizontal drains (see *Figure 9*). Most codes require that a roof area of up to 10,000 square feet have a minimum of two drains. In some installations, a vertical wall diverts rainwater to the roof.

ON THE LEVEL

Combined Sewer Overflows

Some sewer systems are intentionally designed to overflow. The wastewater from combined sewer systems is made up of storm water, domestic sewage, and industrial wastewater. Approximately 900 cities in the United States use combined sewer systems. During periods of heavy rainfall or snowmelt, the flow can exceed the system's capacity. Engineers design the system to drain into a nearby body of water when that happens. The result is called a combined sewer overflow, or CSO.

The Environmental Protection Agency (EPA) issued a control policy for CSOs in 1994. CSOs are classified as urban wet weather discharges. That means CSOs are considered similar to other sanitary sewer overflows and storm water discharges. The EPA's policy on CSOs helps communities meet the discharge and pollutant goals of the Clean Water Act. Take time to review the CSO policy and the Clean Water Act requirements on the EPA's website at www.epa.gov.

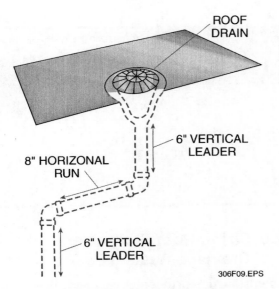

Figure 9 ♦ Typical roof drain in a conventional drainage system.

If so, add half of the wall's area to the projected roof area.

Review the rainfall tables for your local area (refer to *Table 8*). Then convert that figure into gallons per minute per square foot (refer to *Table 10*). Size vertical conductors and leaders and horizontal drains and branches using tables in your local code (see *Tables 11* and *12*). Plumbers usually install horizontal branches with a 1-percent slope. This equals ⅛ inch of slope for every 12 inches of pipe. Increasing the slope will increase the amount of water drained per minute without increasing the pipe size. Refer to your local code before increasing the slope.

To find the total amount of water that a single drain can handle, multiply the number of gallons per square foot per minute by the number of square feet. To calculate roof drain sizes, use the following procedure:

Step 1 Determine the square footage of the roof area. Divide the roof area by 10,000 square feet. This is the number of drains required for the roof.

Step 2 Determine the local rainfall rate and convert to gallons per minute.

Step 3 Multiply the roof area by the number in Step 2. This is the number of gallons per minute that will land on the roof.

Step 4 Divide the number in Step 3 by the number of drains. This is how much storm water each individual drain must handle.

Step 5 Refer to your local code to find the size of the vertical leader and the slope of the horizontal drain.

Step 6 Determine the sizes of the vertical and horizontal pipes into which the individual drains will drain.

The illustration in *Figure 10* shows how to use these steps to size drains for a roof that is 200 feet by 300 feet. In this example, the roof drains have to handle 416 gallons per hour. Based on the information provided, use a 6-inch vertical leader and an 8-inch horizontal drain sloped at ¼ inch per foot.

Table 11 Sample Table for Sizing Vertical Leaders

Diameter of Leader (in.)[a]	Horizontally Projected Roof Area (sq. ft.) Rainfall Rate (in./hr.)											
	1	2	3	4	5	6	7	8	9	10	11	12
2	2,880	1,440	960	720	575	480	410	360	320	290	260	240
3	8,800	4,400	2,930	2,200	1,760	1,470	1,260	1,100	980	880	800	730
4	18,400	9,200	6,130	4,600	3,680	3,070	2,630	2,300	2,045	1,840	1,675	1,530
5	34,600	17,300	11,530	8,650	6,920	5,765	4,945	4,325	3,845	3,460	3,145	2,880
6	54,000	27,000	17,995	13,500	10,800	9,000	7,715	6,750	6,000	5,400	4,910	4,500
8	116,000	58,000	38,660	29,000	23,200	19,315	16,570	14,500	12,890	11,600	10,545	9,600

[a] Sizes indicated are the diameter of circular piping. This table is applicable to piping of other shapes provided the cross-sectional shape fully encloses a circle of the diameter indicated in this table.

Source: International Code Council, Inc.

Table 12 Sample Table for Sizing Horizontal Storm Drains

Size of Piping (in.)	Horizontally Projected Roof Area (sq. ft.) Rainfall Rate (in./hr.)					
	1	2	3	4	5	6
⅛ Unit Vertical in 12 Units Horizontal (1 Percent Slope)						
3	3,288	1,644	1,096	822	657	548
4	7,520	3,760	2,506	1,800	1,504	1,253
5	13,360	6,680	4,453	3,340	2,672	2,227
6	21,400	10,700	7,133	5,350	4,280	3,566
8	46,000	23,000	15,330	11,500	9,200	7,600
10	82,800	41,400	27,600	20,700	16,580	13,800
12	133,200	66,600	44,400	33,300	26,650	22,200
15	218,000	109,000	72,800	59,500	47,600	39,650
¼ Unit Vertical in 12 Units Horizontal (2 Percent Slope)						
3	4,640	2,320	1,546	1,160	928	773
4	10,600	5,300	3,533	2,650	2,120	1,766
5	18,880	9,440	6,293	4,720	3,776	3,146
6	30,200	15,100	10,066	7,550	6,040	5,033
8	65,200	32,600	21,733	16,300	13,040	10,866
10	116,800	58,400	38,950	29,200	23,350	19,450
12	188,000	94,000	62,600	47,000	37,600	31,350
15	336,000	168,000	112,000	84,000	67,250	56,000
½ Unit Vertical in 12 Units Horizontal (4 Percent Slope)						
3	6,576	3,288	2,295	1,644	1,310	1,096
4	15,040	7,520	5,010	3,760	3,010	2,500
5	26,720	13,360	8,900	6,680	5,320	4,450
6	42,800	21,400	13,700	10,700	8,580	7,140
8	92,000	46,000	30,650	23,000	18,400	15,320
10	171,600	85,800	55,200	41,400	33,150	27,600
12	266,400	133,200	88,800	66,600	53,200	44,400
15	476,000	238,000	158,800	119,000	95,300	79,250

Source: International Code Council, Inc.

Step A The roof area equals 300 ft × 200 ft, or 60,000 sq ft.
Step B Using a 4 in/hr rainfall and the information in the sizing table, 0.0416 gal of water will be present on each square foot of roof surface each minute.
Step C 60,000 sq ft of roof area multiplied by 0.0416 gal of water per sq ft equals 2,496 total gal of water falling on the roof each minute.
Step D If 6 drains are required for this installation, then 2,496 gal of water divided by 6 drains equals 416 gpm.
Step E Using the sizing table, 416 gpm requires a 6-in drain (vertical leader).
Step F Using the sizing table, 416 gpm requires an 8-in horizontal drain pipe with a ¼-in pitch.
Step G Using the drainage pattern shown, it can be seen that each pair of roof drains will drain to a larger horizontal drain pipe of 10 in, which carries a total of 832 gpm.
Step H To handle all the water from all 6 drains (2,496 gal), a 15-in horizontal line is needed.

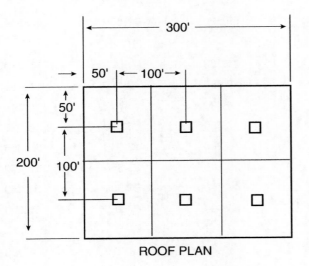
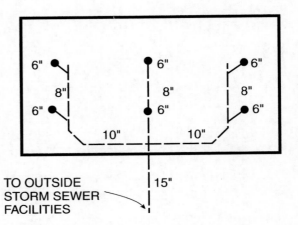

Flow Capacity for Storm Drainage Systems (gpm)				
Pipe Diameter (in)	Roof Drains and Vertical Leaders (gpm)	Horizontal Storm Drainage Piping (gpm)		
		Slope (in/ft)		
		⅛	¼	½
2	30	—	—	—
2½	54	—	—	—
3	92	34	48	69
4	192	78	110	157
5	360	139	197	278
6	563	223	315	446
8	1,208	479	679	958
10	—	863	1,217	1,725
12	—	1,388	1,958	2,775
15	—	2,479	3,500	4,958

Figure 10 ♦ How to size a conventional roof drainage system.

3.2.2 Sizing Controlled-Flow Roof Drainage Systems

Controlled-flow roof drainage systems do not drain water immediately. Instead, the roof acts like a retention pond. Controlled-flow systems hold storm water for a set period of time, usually 12 hours. Then they allow the water to drain slowly through conductors to the main storm drain. This method keeps the storm sewer system from overloading during heavy rainfall.

Controlled-flow systems are also called roof retention systems. They can be installed in one- or two-story buildings if the structure permits. Systems should hold between 3 and 6 inches of storm water. Install overflows in case the rainfall exceeds this amount. The controlled drainage rate means you can install smaller drain piping. Consult your local code for sizing tables for controlled-flow systems (see *Table 13*).

When calculating the amount of water contained by a roof retention system, assume that all the drains are blocked. Your local code will specify how to size and install drains in a roof retention system. In general, use the following steps when installing a controlled-flow roof drainage system:

Step 1 Determine the square footage of the roof area. Divide the roof area by 10,000 square feet. This is the number of drains required for the roof.

Step 2 Determine the slope of the roof and the maximum amount of water to be retained.

Step 3 Determine the local rainfall rate and convert to the peak flow rate.

Step 4 Multiply the square footage of the roof by the peak flow rate. This is the total storm water flow for the roof. It is called the real flow rate.

Step 5 Divide the real flow rate by the number of drains in the system. The result is the real flow per drain.

Step 6 Select the appropriate drain size and the proper slope for the horizontal drain from tables in your local code.

Figure 11 illustrates how to use these steps to size a controlled-flow system for the same 200-foot by 300-foot roof as in *Figure 10*. In this example, use a 3-inch pipe on the more distant drains to remove 31 gallons per minute. The closer drain must handle 62 gallons per minute. Install a 4-inch pipe for these drains. Use a 6-inch line for the total

Table 13 Sample Table for Sizing Controlled-Flow Roof Drainage Systems

Rainfall (in./hr.)	Maximum Storage Depth (in.)	Peak Flow (gpm/sq. ft.)
6.0	3.0	0.06240
5.9	3.0	0.06030
5.8	3.0	0.05820
5.7	3.0	0.05620
5.6	3.0	0.05400
5.5	3.0	0.05020
5.4	3.0	0.05000
5.3	3.0	0.04780
5.2	3.0	0.04580
5.1	3.0	0.04360
5.0	3.0	0.04160
4.9	3.0	0.03960
4.8	3.0	0.03740
4.7	3.0	0.03540
4.6	3.0	0.03320
4.5	3.0	0.03120
4.4	3.0	0.02920
4.3	3.0	0.02700
4.2	3.0	0.02500
4.1	3.0	0.02280
4.0	3.0	0.02080
3.9	3.0	0.01880
3.8	3.0	0.01660
3.7	3.0	0.01460
3.6	3.0	0.01240
3.5	3.0	0.01040
3.4	3.0	0.00832
3.3	3.0	0.00620
3.2	3.0	0.00400
3.1	3.0	0.00320
3.0	2.9	0.00310
2.9	2.8	0.00299
2.8	2.7	0.00288
2.7	2.6	0.00277
2.6	2.5	0.00267
2.5	2.4	0.00256
2.4	2.3	0.00245
2.3	2.2	0.00234
2.2	2.1	0.00223
2.1	2.0	0.00213
2.0	1.9	0.00202
1.9	1.8	0.00192
1.8	1.7	0.00181
1.7	1.6	0.00171
1.6	1.5	0.00160
1.5	1.4	0.00149
1.4	1.3	0.00139
1.3	1.2	0.00128
1.2	1.1	0.00118
1.1	1.0	0.00107
1.0	0.9	0.00097

Step A The roof area equals 300 ft × 200 ft, or 60,000 sq ft.
Step B The roof surface has a normal slope.
Step C The maximum water depth to be retained is 4 in.
Step D The number of drains used is 6.
Step E The rainfall for a given location per hour is 3 in.
Step F Using the sizing table, the peak flow factor is 0.0031 gpm/sq ft.
Step G The total square footage of roof (60,000) times the real flow factor (0.0031) equals the real flow (186 gpm).
Step H The total peak flow (186 gpm) divided by the number of drains (6) equals the real flow per drain (31).
Step I A water storage depth is determined for the given structure. For this building, 4.9 in is used.
Step J The real flow discharge is repeated (31 gpm).
Step K Select a drain size from the chart for a flow rate of 31 gal (3 in).
Step L Select a slope for the horizontal drain line (1⅛" per ft).

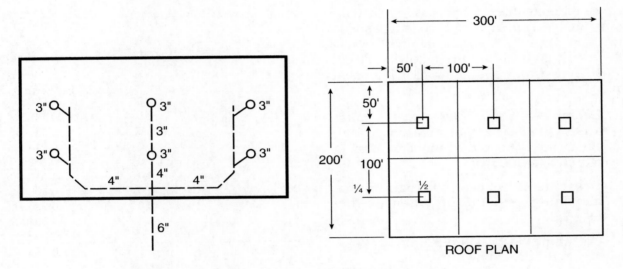

Flow Capacity for Storm Drainage Systems (gpm)				
Pipe Diameter (in)	Roof Drains and Vertical Leaders (gpm)	Horizontal Storm Drainage Piping (gpm)		
		Slope (in/ft)		
		⅛	¼	½
2	30	—	—	—
2½	54	—	—	—
3	92	34	48	69
4	192	78	110	157
5	360	139	197	278
6	563	223	315	446
8	1,208	479	679	958
10	—	863	1,217	1,725
12	—	1,388	1,958	2,775
15	—	2,479	3,500	4,958

Figure 11 ◆ How to size a controlled-flow roof drainage system.

drainage, which is 186 gallons per minute. Compare these dimensions with the previous example. You can see how roof retention systems can drastically reduce the amount of storm water drainage piping. This can mean savings for the client.

Install strainers on all roof drains. If the inlet area is above roof level, size the strainer at least one and a half times the area of the leader. Flat surface-type drain strainers must have an inlet with at least twice the area of the leader. Ensure that all roof drain connections that pass into the interior of the building are watertight. Use approved flashing materials.

3.2.3 Sizing Secondary Roof Drains

Sometimes the main storm drain can become overloaded or blocked. **Secondary drains** provide a backup in such cases. Secondary drains are also called emergency drains. The *2003 International Plumbing Code®* (*IPC*) requires that secondary drains be completely separate from the main roof drain. The *IPC* also requires the drains to be located above grade and where occupants can see them. If your local code is based on the *2003 IPC,* you can size the secondary drains using the same rainfall rates as for the main roof. Do not consider the flow through the main drain when sizing the secondary drain. Refer to your local code for standards that apply in your area.

Another form of overflow protection is the installation of **scuppers** in the walls or parapets around the roof (see *Figure 12*). These are openings that allow excess water to drain from the roof into the surface drainage system. Refer to your local code for the sizing and placement of scuppers.

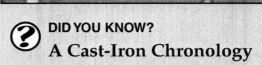

DID YOU KNOW?
A Cast-Iron Chronology

In 1562, a fountain in Langensalza, Germany, was fitted with cast-iron water supply pipe. This is the earliest recorded use of cast-iron pipe. The famous water gardens of Versailles, France, built in 1664, also used cast-iron pipe. These pipes, which still work, carry water from a river 15 miles away. London received its first cast-iron pipe water distribution system in 1746. Sir Thomas Simpson invented the bell-and-spigot joint in 1785. This was an improvement over bolted flanges and butt joints wrapped with metal bands. Bell-and-spigot joints are still widely used in cast-iron pipe.

Around the beginning of the 19th century, cities in Pennsylvania and other eastern states began using cast-iron pipe. At the time, iron ore mined in the United States was shipped to foundries in England, and the completed pipes were shipped back across the ocean. The first successful U.S. foundries were not established until 1819 in New Jersey. Beginning in the 1880s, foundries moved outward to the South and the Midwest, and their pipe plumbed the country's rapidly growing cities. Cast-iron pipe production grew to an industrial scale during that time. Production of cast-iron pipe in the United States peaked at 280,000 net tons in 1916.

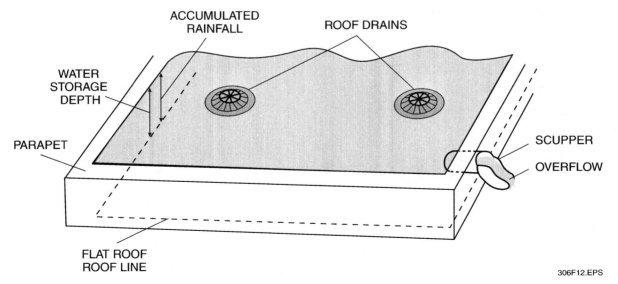

Figure 12 ◆ Scupper installation in a controlled-flow drainage system.

3.3.0 Sizing Above-Grade and Below-Grade Drainage Systems

In some areas, storm water can be allowed to drain directly onto the ground. This action is called **ponding**. The water then **percolates**, or seeps, into the soil. Ponding is permitted only on ground that has been set aside for that purpose. To construct effective aboveground storage, you need to know how well the soil can diffuse the water that flows into it. The ability of soil to absorb water is called **permeability**. Consult your local agricultural extension service to learn about the permeability of soil in your area.

Your local code will list the different types of pipe that can be used in aboveground storm drainage systems. The *IPC* allows the following pipe materials:

- Acrylonitrile butadiene styrene (ABS) plastic
- Brass
- Cast iron
- Copper and copper alloy
- Types K, L, M, and DWV copper and copper-alloy tubing
- Galvanized steel
- Glass
- Polyolefin
- Type DWV polyvinyl chloride (PVC) plastic

In some cases, the design or location of the structure will not allow the ponding of storm water. In those cases, an underground method may be installed. This usually takes the form of an underground retention basin. If the basin is below the sewer grade, you will have to install a pump. You will learn how to size storm water sumps in the module *Sewage Pumps and Sump Pumps*. Acceptable pipe materials for underground storm water drains include:

- ABS plastic
- Cast iron
- Cement or concrete
- Corrugated galvanized steel
- Types K and L copper or copper-alloy tubing
- Polyolefin
- Type DWV PVC plastic

Review Questions

Section 3.0.0

1. Always install a(n) _____ when connecting a storm drain to a combined sewer line.
   a. strainer
   b. trap
   c. sanitary ell
   d. interceptor

2. Never _____ a drainage pipe in the direction of flow.
   a. increase the size of
   b. add branches to
   c. redirect
   d. reduce the size of

3. Most codes require that roof area of up to _____ square feet must have a minimum of two drains.
   a. 10,000
   b. 25,000
   c. 35,000
   d. 60,000

4. A 1 percent slope equals _____ inch of slope for every 12 inches of pipe.
   a. ⅛
   b. ¼
   c. ½
   d. 1

5. Controlled-flow systems usually hold storm water for _____ hours before allowing it to drain slowly through conductors to the main storm drain.
   a. 8
   b. 10
   c. 12
   d. 14

6. The action of water seeping into the soil is called _____.
   a. ponding
   b. percolation
   c. permeation
   d. porosion

Summary

DWV systems and storm drainage systems provide safe and efficient drainage for residences and commercial buildings. Plumbers size both types of system, and codes vary with regard to drainage requirements.

This module addressed the various techniques that plumbers use to size DWV systems. Plumbers size DWV systems by calculating the DFU load in each branch. DFUs measure the waste discharge of fixtures. Local codes provide tables for using DFU loads to size building drains and sewers. Vents are also sized by calculating the DFUs in the system. Plumbers use the same techniques to size combination drain and vent systems and reduced size vent systems.

This module also discussed methods to size a storm drainage system. Plumbers consider the maximum expected rainfall and the rate of removal. Local codes provide rainfall tables, which are used to estimate the system's capacity. Above- and below-grade systems allow storm water to percolate into the soil.

The type of roof drainage system affects its sizing. Conventional systems allow storm water to run off as soon as it falls. Controlled-flow systems retain water on the roof and release it slowly, allowing for smaller piping. Local codes provide sizing tables for both types of roof drainage systems.

Notes

Trade Terms Introduced in This Module

Conductor: A vertical spout or pipe that drains storm water into a main drain. It is also called a leader.

Controlled-flow roof drainage system: A roof drainage system that retains storm water for an average of 12 hours before draining it. Controlled-flow systems help prevent overloads in the storm-sewer system during heavy storms.

Conventional roof drainage system: A roof drainage system that drains storm water as soon as it falls.

Drainage fixture unit (DFU): A measure of the waste discharge of a fixture in gallons per minute. DFUs are calculated from the fixture's rate of discharge, the duration of a single discharge, and the average time between discharges.

Percolate: To seep into the soil and become absorbed by it.

Permeability: A measure of the soil's ability to absorb water that percolates into it.

Ponding: Collecting of storm water on the ground.

Scupper: An opening on a roof or parapet that allows excess water to empty into the surface drain system.

Secondary drain: A drain system that acts as a backup to the primary drain system. It is also called an emergency drain.

Resources & Acknowledgments

Additional Resources

This module is intended to be a thorough resource for task training. The following reference works are suggested for further study. These are optional materials for continued education rather than for task training.

Code Check Plumbing: A Field Guide to Plumbing, Second Edition. 2004. Michael Casey, Redwood Kardon, and Douglas Hansen. Newtown, CT: Taunton Press.

Planning Drain, Waste & Vent Systems. 1990. Howard C. Massey. Carlsbad, CA: Craftsman Book Company.

Water, Sanitary, and Waste Services for Buildings. 2002. Alan F. E. Wise and J. A. Swaffield. Boston, MA: Addison-Wesley.

References

Dictionary of Architecture and Construction, Third Edition. 2000. Cyril M. Harris, ed. New York: McGraw-Hill.

Pipefitters Handbook. 1967. Forrest R. Lindsey. New York: Industrial Press Inc.

2003 International Plumbing Code. 2003. International Code Council. Falls Church, VA: International Code Council.

Cast Iron Soil Pipe Institute website, www.cispi.org, *Cast Iron Soil Pipe & Fittings Handbook,* http://www.cispi.org/handbook.htm, reviewed June 2005.

Efficient Building Design, Volume III, Water and Plumbing. 2000. Ifte Choudhury and J. Trost. Upper Saddle River, NJ: Prentice-Hall.

Figure Credits

Menzies Metal Products, 306F01

Hydrometeorological Design Studies Center, National Weather Service, NOAA, Table 8

International Code Council, Inc., *2000 IPC Commentary,* 306F05, 306F06, 306F08, Tables 1 through 7, 11, 12

CONTREN® LEARNING SERIES — USER UPDATE

The NCCER makes every effort to keep these textbooks up-to-date and free of technical errors. We appreciate your help in this process. If you have an idea for improving this textbook, or if you find an error, a typographical mistake, or an inaccuracy in NCCER's Contren® textbooks, please write us, using this form or a photocopy. Be sure to include the exact module number, page number, a detailed description, and the correction, if applicable. Your input will be brought to the attention of the Technical Review Committee. Thank you for your assistance.

Instructors – If you found that additional materials were necessary in order to teach this module effectively, please let us know so that we may include them in the Equipment/Materials list in the Annotated Instructor's Guide.

Write: Product Development and Revision
National Center for Construction Education and Research
P.O. Box 141104, Gainesville, FL 32614-1104

Fax: 352-334-0932

E-mail: curriculum@nccer.org

Craft _____ Module Name _____

Copyright Date _____ Module Number _____ Page Number(s) _____

Description
_____
_____
_____
_____

(Optional) Correction
_____
_____
_____

(Optional) Your Name and Address
_____
_____
_____

Plumbing Level Three

02307-06

Sewage Pumps and Sump Pumps

02307-06
Sewage Pumps and Sump Pumps

Topics to be presented in this module include:

1.0.0 Introduction 7.2
2.0.0 Sewage Removal Systems 7.2
3.0.0 Storm Water Removal Systems 7.10
4.0.0 Troubleshooting and Repairing Sewage and
Storm Water Removal Systems 7.13
5.0.0 The Worksheet 7.17

Overview

Building designs often require plumbers to locate a drain below the sewer line. Because these drains, called sub-drains, cannot discharge into the sewer by gravity, the wastewater must be pumped upward to the higher-level drainage lines. Sub-drains empty into temporary holding pits called sumps, where pumps, or ejectors, provide the lift needed. These sewage removal systems, together with storm water removal systems, perform a critical role in plumbing installations and in maintaining public health.

Sewage removal systems use either a centrifugal pump or a pneumatic ejector-type pump, each of which is designed for specific applications. Plumbers should review the design of the building's waste system to ensure they select the most appropriate pump. It may be advisable to install a backup pump in case of lead pump failure; local codes may also require power-off alarms and other precautions for power failures. Storm water removal systems are used mainly for flood control, and while similar to sewage removal systems in design, have very different applications and requirements. Storm water sumps collect runoff from sub-drains and store it until it can be pumped to storm sewers or to flood-control areas.

Quick and efficient troubleshooting of malfunctioning pumps is essential. Typically, pump problems can be traced to either electrical causes (involving the pump motor and wiring) or mechanical causes (involving the pump's moving parts). If a pump needs to be replaced, plumbers must review the relevant drawings and specifications to ensure that the installed pump matches the job it is to perform.

Focus Statement
The goal of the plumber is to protect the health, safety, and comfort of the nation job by job.

Code Note
Codes vary among jurisdictions. Because of the variations in code, consult the applicable code whenever regulations are in question. Referring to an incorrect set of codes can cause as much trouble as failing to reference codes altogether. Obtain, review, and familiarize yourself with your local adopted code.

Portions of this publication reproduce tables and figures from the *2003 International Plumbing Code* and the *2000 IPC Commentary*, International Code Council, Inc., Falls Church, Virginia. Reproduced with permission. All rights reserved.

Objectives

When you have completed this module, you will be able to do the following:

1. Explain the functions, components, and operation of sewage and sump pumps.
2. Size a storm water sump by calculating the runoff from paved and unpaved land surfaces.
3. Size a sewage sump by calculating the sewage flow from a structure.
4. Install and adjust sensors, switches, and alarms in sewage and sump pumps.
5. Troubleshoot and repair sewage and sump pumps.
6. Using a detailed drawing, identify system components.
7. Install a sump pump.
8. Find local applicable code requirements for installation and use.

Trade Terms

Air lock
Centrifugal pump
Drainage fixture unit
Duplex pump
Ejector
Float switch
Impeller
Lift station
Mercury float switch
Pneumatic ejector
Pressure float switch
Probe switch
Retention pond
Retention tank
Reverse flow pump
Sewage pump
Sewage removal system
Storm water removal system
Sub-drain
Sump
Sump pump

Required Trainee Materials

1. Appropriate personal protective equipment
2. Sharpened pencils and paper
3. Copy of local applicable code
4. Calculator

Prerequisites

Before you begin this module, it is recommended that you successfully complete *Core Curriculum; Plumbing Level One; Plumbing Level Two; Plumbing Level Three*, Modules 02301-06 through 02306-06.

This course map shows all of the modules in the third level of the *Plumbing* curriculum. The suggested training order begins at the bottom and proceeds up. Skill levels increase as you advance on the course map. The local Training Program Sponsor may adjust the training order.

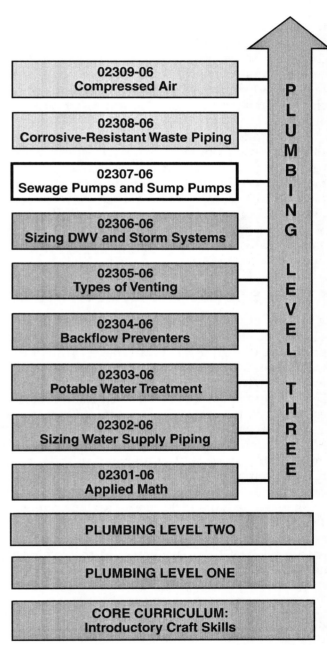

1.0.0 ♦ INTRODUCTION

Plumbing systems conduct wastewater away for disposal through drains. Drains prevent contamination by sewage wastes and flooding by runoff water. You learned how to size and install drains earlier in this curriculum. Often, the design of a building makes it necessary for a plumber to locate a drain below the sewer line. These drains are called **sub-drains**. Such drains can't discharge into the sewer by gravity. Instead, sub-drains must empty into temporary holding pits. Pumps installed in the pits provide the lift to move wastewater up out of the pit and into higher drainage lines. Pits and pumps that handle sewage wastes are called **sewage removal systems** or **lift stations**. Pits and pumps that handle clear water runoff are called **storm water removal systems**.

This module reviews the components that plumbers use to build sewage and storm water removal systems. It also discusses how to troubleshoot and repair the most common problems with pumps, controls, and pits. Always review your local code before designing, installing, or repairing pumps and other components. Sewage and storm water removal systems help maintain public health and safety.

2.0.0 ♦ SEWAGE REMOVAL SYSTEMS

A complete sewage removal system consists of **sewage pumps**, **sumps**, and controls. A sewage pump creates a partial vacuum that draws the waste from the pit into the pump and forces it out with sufficient pressure to reach the sewer above. Sewage pumps are often called **ejectors**. A sump collects sewage from the sub-drain. Sumps are also called wet wells or sump pits. Each installation has controls that operate the pump. Controls measure the amount of wastewater in the sump, turn the pump on and off, and provide backup in case of pump or power failure.

There are two distinct types of sewage removal systems: centrifugal and pneumatic. The names refer to the type of pump used to remove the waste. Each of these systems and their components are discussed in more detail in the following sections.

2.1.0 Sewage Pumps

Two types of pumps are used in sewage removal systems: **centrifugal pumps** and **pneumatic ejectors**. Both types of pump are designed for specific applications. Be sure to review the design of the building's waste system. This will help you select the most appropriate type of pump. Always follow the manufacturer's instructions when installing pumps and ejectors. Join pumps to the inlet and outlet piping with either a union or a flange. This permits the pump to be easily removed for maintenance, repair, or replacement. The following installation steps are common for most sewage and storm water pumps:

Step 1 Before installing the pump in the basin, ensure the pump is appropriate for the sump.

Step 2 Check that the voltage and phase from the electrical supply matches the power requirements shown on the pump's nameplate.

Step 3 Thread the sump's discharge pipe into the pump's discharge connection.

Step 4 Install a check valve horizontally if one is required. If this is not possible, install it at a 45-degree angle with the pivot at the vertical to prevent solids from clogging the valve.

Step 5 Drill a 3/16-inch (or recommended size) hole in the discharge pipe about 2 inches above the pump discharge connection to prevent air from locking the pump.

Step 6 If required, install a gate valve in the system after the check valve to permit removal of the pump for servicing.

Step 7 Install a union above the high-water line between the check valve and the pump to allow the pump to be removed without disturbing the piping.

2.1.1 Centrifugal Pumps

Centrifugal pumps are frequently used in both storm water and sewage removal systems. Centrifugal pumps consist of the pump, an electric motor to run the pump, a plate on which to mount the pump, and controls to turn the pump on and off at the proper time. Many pumps come with motors that are designed to be submersible, which means the entire pump assembly can be placed inside the sump (see *Figure 1*). Others have a motor that must be installed above the sump, with the pump body connected to the motor by a sealed shaft (see *Figure 2*). Centrifugal pumps are widely used in sewage removal systems because of their mechanical simplicity and low cost. Complete centrifugal pump systems are available as self-contained units with pump, controls, and basin. They may also be fabricated on site from individual components.

DID YOU KNOW?
History of the Pump

One of the earliest known pumps was the shaduf, developed in Egypt around 1550 B.C.E. The shaduf was a lever that lifted water from a well. About 400 years later in Persia, people pumped water using a chain of pots tied to a long loop of rope. The loop was wrapped around a wheel. As the wheel turned, the pots scooped water from a well and brought it to the top. About 650 B.C.E., Romans in Egypt pumped water with a similar device called a noria. This was a wheel with pots attached to the rim. When the wheel was lowered partly into the water and spun, the buckets dipped into the water. In 230 B.C.E., the Greek inventor Archimedes designed a pump made out of a coil of pipe wrapped around a shaft. By dipping one end of the pipe into the water and turning the shaft, the water corkscrewed its way up the pipe coil and came out the top end. Water wheels and Archimedean screws are still widely used throughout the world.

The Byzantines used a piston and cylinder to lift water using the force of air pressure in the first century B.C.E. Air pressure is also used in the centrifugal pump, which was invented by Leonardo da Vinci in the 1500s. It employs a set of spinning blades to force water into and out of a chamber. An engineer named Denis Papin designed the first practical centrifugal pump in 1688. The centrifugal pump is widely used in modern sewage and storm water removal systems.

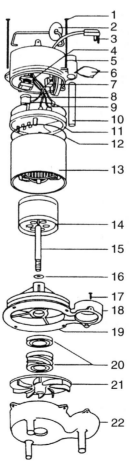

1	STAINLESS STEEL BOLTS
2	CARRYING HANDLE AND DATE TAG
3	POWER CORD
4	MOTOR START RELAY
5	MOTOR COVER
6	HUMISORB PACKET
7	WATER LEVEL SWITCH
8	OVERLOAD PROTECTOR
9	CONNECTING LEADS
10	PVC-AIR TUBE
11	BAFFLE AND TERMINAL PLATE
12	RUBBER GASKET
13	STATOR ASSEMBLY – 115V
14	ROTOR SHAFT ASSEMBLY
15	SHAFT
16	THRUST RACE
17	SCREW
18	CAST-IRON MOTOR END
19	DIVERTER FLANGE
20	CERAMIC SEAT AND ROTARY SEAL HEAD
21	IMPELLER
22	HOUSING PLATE

Figure 1 ◆ Submersible centrifugal sewage pump.

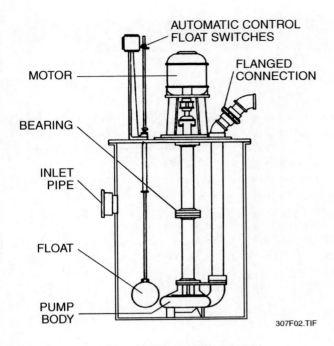

Figure 2 ◆ Centrifugal pump installation with aboveground motor.

Inside the body of a centrifugal pump is a set of spinning vanes called an **impeller**. An impeller removes wastewater from a sump in two steps. First, it draws wastewater up from the sump by creating a partial vacuum. It then ejects the swirling water out of the pump body into the discharge line. Centrifugal pumps are water-lubricated, which means that they must always be in contact with water or they will suffer damage. If this happens, repair or replace the pump. Ensure that centrifugal pumps are primed before switching them on.

Centrifugal pumps can be designed with nonclogging and self-grinding features that efficiently handle solid wastes up to a certain size. The *International Plumbing Code® (IPC)*, for example, requires that pumps receiving wastes from water closets must be able to handle solids up to 2 inches in diameter. Other pumps must be able to pump solids up to 1 inch in diameter. Because sand and grit in the wastewater can cause significant wear on pump components, install sediment interceptors in the drain line. Centrifugal pumps may require maintenance to replace moving parts like the impeller. Install pumps so that they can be removed easily from sumps.

Sometimes a design may call for a backup capability in case a pump fails, or a design may require extra pumping capacity during periods of peak flow. In such cases, install a **duplex pump** to fulfill the requirement. A duplex pump consists of two complete centrifugal pump assemblies, including floor plates, controls, and high-water alarms, installed in parallel in a sump (see *Figure 3*). Duplex pumps can be designed so that they operate on an alternating basis, reducing the wear on a single pump. Use an electrical pump alternator to switch between pumps in a duplex arrangement. Remember to size the sump to hold both pumps. Sump covers must be specially ordered to accommodate a duplex installation.

Rags and stringy solids may clog centrifugal pumps. Floating wastes and scum are out of the reach of a pump inlet that is located at the bottom of a sump. For these types of sewage wastes, use a **reverse flow pump** arrangement. Reverse flow pumps consist of a duplex pump installed with strainers, cutoff valves, and check valves (see *Figure 4*). Reverse flow pumps are designed to discharge solid wastes without passing them through the impeller housing. Because reverse flow pumps consist of two separate pumps, the system offers a safety backup in case one pump fails.

As wastewater enters a sump through one inlet, a strainer traps solids. When the water reaches a predetermined height, the pump switches on. The pump forces the trapped solids out of the strainer and through the discharge pipe. To prevent back pressure into the soil line when the pump is operating, install a full-opening, positive-seating, noncorrosive check valve. Locate the check valve in the horizontal drain line (refer to *Figure 4*). The duplex arrangement allows wastewater to flow into the basin through one of the inlets while the other is closed during the pump discharge process. Install a pump alternator to switch between the two pumps. Occasionally, excess amounts of wastewater may surge into the basin, requiring both pumps to activate at the same time. To allow for this, install overflow strainers near the inlets for each pump. Overflow strainers allow wastewater to enter the sump when both check valves are closed.

2.1.2 Pneumatic Ejectors

A pneumatic ejector is an alternative to a centrifugal pump. Plumbers often select pneumatic ejectors for large-capacity applications, such as housing subdivisions and municipal facilities. Unlike centrifugal pumps, pneumatic ejectors do not have moving parts. This means they require less maintenance than centrifugal pumps. However, they are more mechanically complex than centrifugal systems.

Sewage enters a pneumatic ejector through an inlet. When the wastes reach a predetermined height, a sensor triggers a compressor. The compressor forces air into the basin. The increased pressure causes the sewage to be moved into the discharge line and out of the basin. Pneumatic

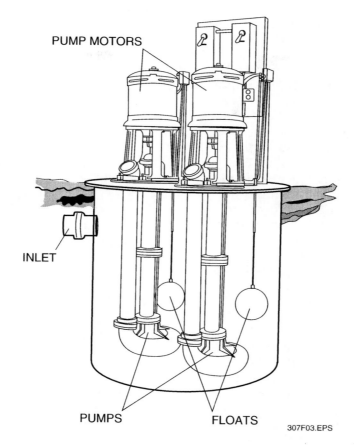

Figure 3 ♦ Duplex pump installation.

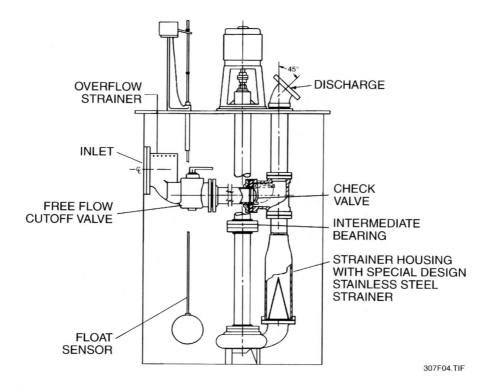

Figure 4 ♦ Reverse flow pump installation.

ejectors are equipped with check valves on the inlet and discharge lines. During waste ejection, the valves prevent back pressure. After ejection, they keep the waste from backflowing. When the ejector is empty, a diaphragm exhaust valve releases the compressed air and allows the chamber to refill with sewage (see *Figure 5*).

Pneumatic ejectors are designed as a unit. This means that basins, pumps, and controls are all part of a single manufactured package. Be sure to install the right size pneumatic ejector for the anticipated load. Consult the construction drawings, and check your local code for requirements governing pneumatic ejector installation. Install a union above the high-water line between the check valve and the pump to allow the pump to be removed for maintenance without disturbing the piping.

2.2.0 Sewage Sumps

A sump is a container that holds wastewater until it can be pumped into the sewer line. Plumbers are responsible for installing sumps. Installation includes all of the components within the sump, including pumps, controls, and inlet and outlet piping. Indoor sumps are made of plastic or fiberglass (see *Figure 6*). Outdoor sumps can be precast or site-built from concrete (see *Figure 7*). Basin size and materials are specified in your local code. On pneumatic systems, install check valves on sump inlet piping, and install gate and check valves on discharge piping. Ensure that the valves are located outside the sump and are easily accessible. Residential installations may require only a check valve; refer to your local code.

WARNING!
Wear appropriate personal protective equipment when inspecting sewage pumps. Review the manufacturer's specifications before inspecting or removing a sewage pump.

When designing sumps, remember to consider the drain flow rate from all of the fixtures draining into the sub-drain. Calculate the flow in terms of the **drainage fixture unit**. A drainage fixture unit is a measure of a fixture's discharge in gallons per minute divided by 7.5. This is the number of gallons in a cubic foot. Drainage fixture unit calculations take several factors into consideration:

- Rate of discharge
- Duration of discharge
- Average time between discharges

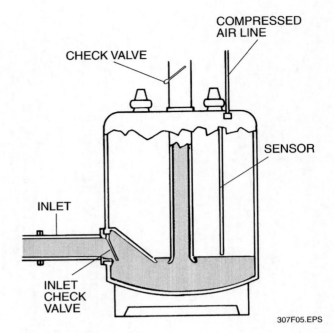

Figure 5 ◆ Pneumatic ejector.

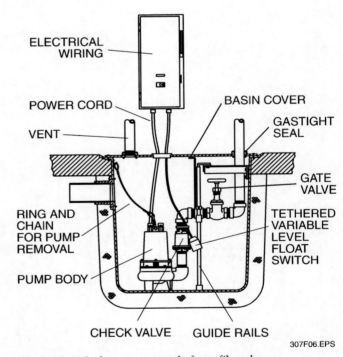

Figure 6 ◆ Indoor sump made from fiberglass.

Table 1 shows drainage fixture units for fixture drains or traps, and *Table 2* provides drainage fixture units for fixtures in commercial and private installations.

Install airtight covers and vents on sewage sumps. Otherwise, sewer gases will escape and may pose a threat to health and safety. Consult local codes for specific venting requirements. Design the sewage removal system so that wastes are not retained for more than 12 hours. The maximum

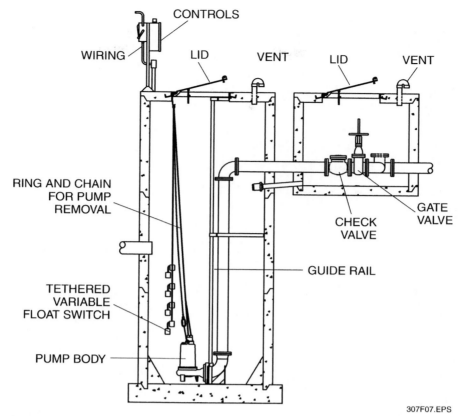

Figure 7 ♦ Outdoor sump basin made from concrete.

Table 1 Sample Drainage Fixture Units for Fixture Drains or Traps

Fixture Drain or Trap Size (inches)	Drainage Fixture Unit Value
1¼	1
1½	2
2	3
2½	4
3	5
4	6

Source: International Code Council, Inc.

allowable height of wastewater in sumps is established in local codes. Ensure that the pump does not cycle more than six times per hour because excessive cycling could damage the pump.

2.3.0 Controls

When wastewater in a sump reaches a preselected height, a switch in the basin activates the pump. Switches are also called sensors. Sewage removal systems use any one of four types of switches:

- **Float switch**
- **Pressure float switch**
- **Mercury float switch**
- **Probe switch**

A float switch works a lot like the float in a water closet tank. As the wastewater level rises, the float rises within the sump. An arm attached to the float activates the pump when the float reaches a specified level.

A pressure float switch is an adjustable type of float switch. Install pressure float switches to rise with wastewater or clamp them to the basin wall (see *Figure 8*).

A mercury float switch is a specialized float sensor that plumbers can install at various heights within the basin (see *Figure 9*). Usually two mercury

Table 2 Sample Drainage Fixture Units for Fixtures and Groups

Fixture Type	DFU Value as Load Factors	Minimum Diameter of Trap (in.)
Automatic clothes washers, commercial[a,g]	3	2
Automatic clothes washers, residential[g]	2	2
Bathroom group (water closet, lavatory, bathtub or shower, with or without a bidet, and an emergency drain), 1.6 gal per flush water closet[f]	5	—
Bathroom group (as above), >1.6 gal per flush water closet[f]	6	—
Bathtub[b] (with or without overhead shower or whirlpool attachments)	2	1½
Bidet	1	1¼
Combination sink and tray	2	1½
Dental lavatory	1	1¼
Dental unit or cuspidor	1	1¼
Dishwashing machine[c], domestic	2	1½
Drinking fountain	½	1¼
Emergency floor drain	0	2
Floor drains	2	2
Kitchen sink, domestic	2	1½
Kitchen sink, domestic, with food waste grinder and/or dishwasher	2	1½
Laundry tray (1 or 2 compartments)	2	1½
Lavatory	1	1¼
Shower	2	1½
Sink	2	1½
Urinal	4	(note d)
Urinal, ≤1gal per flush	2[e]	(note d)
Wash sink (circular or multiple), each set of faucets	2	1½
Water closet, flushometer tank, public or private	4[e]	(note d)
Water closet, private (1.6 gal per flush)	3[e]	(note d)
Water closet, private (>1.6 gal per flush)	4[e]	(note d)
Water closet, public (1.6 gal per flush)	4[e]	(note d)
Water closet, public (>1.6 gal per flush)	6[e]	(note d)

[a] This is not applicable for traps larger than 3 inches.
[b] A shower head over a bathtub or whirlpool bathtub attachment does not increase the DFU value.
[c] See local code for methods of computing unit value of fixtures not listed or for rating of devices with intermittent flows.
[d] Trap size shall be consistent with fixture outlet size.
[e] For the purpose of computing loads on building drains and sewers, water closets and urinals shall not be rated at a lower DFU unless the lower values are confirmed by testing.
[f] For fixtures added to a dwelling unit bathroom group, add the DFU value of those additional fixtures to the bathroom group fixture count.
[g] See your local code for sizing requirements for fixture drain, branch drain, and drainage stack for an automatic clothes washer standpipe.

Source: International Code Council, Inc.

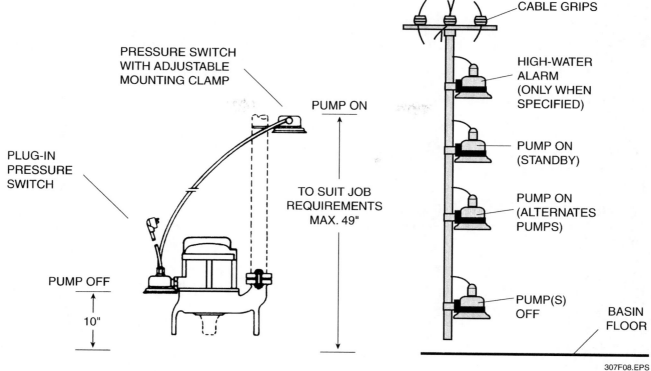

Figure 8 ◆ Pressure float switches in a sewage sump basin.

float switches are required, and a third switch can serve as a high-water alarm if desired. If more than one pump is used in the basin, install four mercury float switches.

Probe switches, also called electrodes, are used in pneumatic ejectors. They work similarly to thermostatic probes installed on water heaters. When wastewater in the sump reaches a certain height, the probe triggers the compressor. When the waste is ejected, the switch resets.

Sewage removal systems perform a critical role in plumbing installations. Proper sizing and installation will help ensure that a system will not suffer from malfunctions and breakdowns. When possible, provide a backup pump in case the lead pump fails. For pneumatic ejector installations, provide a backup compressor. Power failures will prevent centrifugal pumps and pneumatic ejectors from working. Your local applicable code may require the installation of power-off alarms for power failures and similar conditions. If possible, provide a backup power supply. Remember that plumbers are responsible for helping to maintain public safety through the fast and efficient removal of wastes. These extra steps will allow the system to continue operating while plumbers are called in to make repairs.

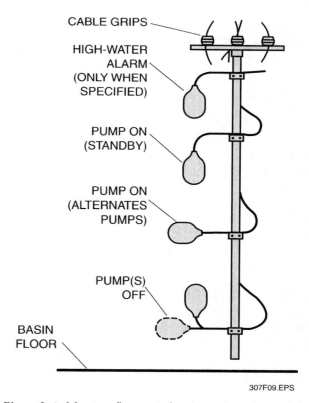

Figure 9 ◆ Mercury float switches in a sewage sump basin.

Review Questions

Sections 1.0.0–2.0.0

1. If it is not possible to install a check valve horizontally, install it _____.
   a. at a 45-degree angle
   b. vertically
   c. at a 60-degree angle
   d. parallel

2. To allow a pump to be removed without moving the piping, install a union above the high-water line between the _____ and the pump.
   a. discharge pipe
   b. inlet pipe
   c. gate valve
   d. check valve

3. _____ allow wastewater to enter a sump when the check valves are closed.
   a. Gate valves
   b. Float switches
   c. Overflow strainers
   d. Cutoff valves

4. Calculate drainage fixture units by dividing a fixture's discharge in gallons per minute by _____.
   a. 0.0133
   b. 3.14
   c. 7.5
   d. 8.3

5. When installing a mercury float switch, usually _____ switches are required.
   a. four
   b. two
   c. eight
   d. six

3.0.0 ◆ STORM WATER REMOVAL SYSTEMS

Roofs, paved areas, and yards all require drains. The water from these areas drains into a separate storm sewer system or into special flood control areas, such as ponds and tanks. Many codes do not allow storm water to be discharged into sewage lines. Install a storm sewer line to drain clear water runoff. If the collection point for runoff must be deeper than the sewer drain line, install a storm water removal system to lift the wastewater.

Storm water removal systems are mainly used for flood control. The procedures for sizing and installing storm water removal systems share some similarities with sewage removal systems. However, keep in mind that the applications of the two types of systems are very different. Storm water pumps and basins are designed for clear water wastes rather than wastes with solid matter. Local codes provide guidance on how to design storm water removal systems.

3.1.0 Storm Water Pumps

Pumps used in storm water removal systems are often called **sump pumps** or bilge pumps. Install centrifugal sump pumps to lift storm water from sumps to the drainage lines (see *Figure 10*). Like centrifugal sewage pumps, the motor of a centrifugal sump pump may be mounted either atop or inside the sump. Install pumps made of chlorinated polyvinyl chloride (CPVC) where the runoff contains corrosive chemicals (see *Figure 11*).

Many local codes require gate and check valves on the discharge line (see *Figure 12*). Check valves prevent backflow into the basin. Aluminum flapper check valves are often specified because of their quieter operation. In a single or duplex pump, the check valve should be installed in a horizontal section of the discharge line. Gate valves, which allow the plumber to shut off the discharge for maintenance, may be placed in either a horizontal or a vertical run of the discharge line.

DID YOU KNOW?
Pumps in History

Pumps have played a vital supporting role in some of history's greatest inventions and scientific discoveries. For example, in 1660, English chemist Robert Boyle (1627–1691) published a book called *The Spring and Weight of the Air* in which he used a hand pump of his own design to remove the air from a large sealed chamber. He placed a bell in the chamber and showed how the sound of the bell faded as more air was pumped out. He also placed a burning candle in the chamber and showed how it snuffed out as air was pumped out. His experiments with air pressure shattered many widely held beliefs about the nature of air. His work encouraged others to take up scientific experiments. Boyle's use of pumps to discover the nature of air is considered one of the most important examples of how empirical, or experimental, science can lead to new discoveries.

As a young man, the Scottish inventor and engineer James Watt (1736–1819) became interested in steam-powered pumps. Miners used primitive and inefficient steam-powered piston pumps to suck water out of England's coal mines. Watt did research to find ways to improve how the steam pumps worked. His research led him to discover the properties of steam, which in turn led to an improved steam engine design that eventually powered the Industrial Revolution in the nineteenth century.

The next time you install a sewage pump, remember that that humble pump has a noble pedigree!

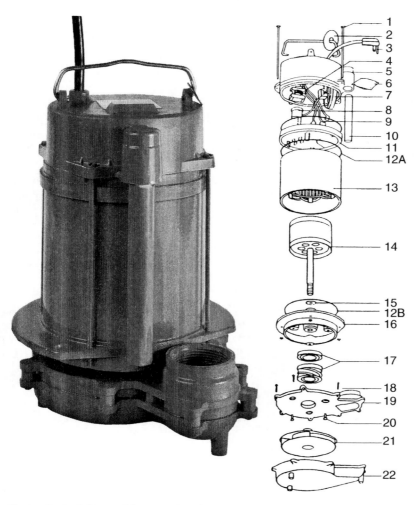

1	STAINLESS STEEL BOLTS
2	CARRYING HANDLE AND DATE TAG
3	POWER CORD
4	MOTOR START RELAY
5	MOTOR COVER
6	HUMISORB PACKET
7	WATER LEVEL SWITCH
8	OVERLOAD PROTECTOR
9	CONNECTING LEADS
10	PVC-AIR TUBE
11	BAFFLE AND TERMINAL PLATE
12A	RUBBER GASKET #050
12B	RUBBER GASKET #160
13	STATOR ASSEMBLY — ½ HP — 115V 60C 7½ A CONTINUOUS DUTY OIL DUROL CC
14	ROTOR SHAFT ASSEMBLY
15	THRUST RACE
16	MOTOR END
17	CERAMIC SEAT AND ROTARY SEAL HEAD
18	STAINLESS STEEL SCREWS TYPE B
19	PUMP VOLUTE
20	FLAT-HEAD SCREWS
21	REINFORCED POLYPROPYLENE IMPELLER
22	HOUSING PLATE

Figure 10 ◆ Submersible centrifugal sump pump.

Size the pump according to the anticipated amount of drainage. You will learn how to calculate rainfall rates later in this module. Provide a safety factor by ensuring that the pump has a capacity of 1½ times the maximum calculated flow. An additional pump can serve as a backup in case the primary pump fails. Power failures often occur during storms, which is when sump pumps are most needed. Consider installing a backup power source to keep the sump pumps running.

3.2.0 Storm Water Sumps

Storm water sumps collect runoff from sub-drains and store it until it can be pumped to storm sewers or to flood-control areas. Construct sumps from cast iron, fiberglass, or concrete. Apply waterproofing to the inside of concrete sumps. Provide a settling area inside sumps. This will allow sand and other coarse material to settle without being sucked up into the pump. Sediment can damage the pump. Sump covers do not need to be airtight; however, a secure lid will prevent objects from falling into the sump.

Sizing sumps properly will ensure that they perform their function efficiently. Sizing a storm water sump is similar to the process for sizing a sewage sump. Keep two objectives in mind when sizing a sump:

- The basin must be large enough to receive the anticipated amount of storm water.
- The basin must be large enough to eliminate frequent cycling of the pump motor.

Ensure that the basin's high-water mark is at least 3 inches below the sub-drain inlet. The low-level mark must be no less than 6 inches above the bottom of the basin. This will ensure that the pump's suction end is always underwater.

Size the sump so that the pump motor does not activate more than once every 5 minutes. For example, if wastewater flows into a sump at 60 gallons per minute (gpm) and if the pump cycles every 5 minutes, then the basin has to hold at least 300 gallons (5 × 60) of wastewater.

How big will a sump have to be to handle 300 gallons of wastewater? You can size basins by using the formula for finding the volume of a cylinder. The formula is $V = (\pi r^2)h$

In other words, the volume is the product of the area of the basin's floor and its height. When converting the volume to gallons, remember that there are 7.5 gallons in a cubic foot. Using the formula, a 4-foot diameter cylinder would have to be 3.18 feet, or 3 feet 2 inches tall, to hold 300 gallons. Adding the required 3-inch minimum at the top and 6-inch minimum at the bottom, the sump should be at least 3 feet 11 inches from the bottom of the inlet pipe to the base of the sump.

Local codes will provide the rate of wastewater collection and area rainfall rates to use when calculating the size of the removal system. Unless otherwise specified, use a collection rate of 2 gpm for every 100 square feet of sandy soil and 1 gpm for clay soils. Flow rates for roof and driveway runoff will vary considerably. For roofs and driveways, use a collection rate of 1 gpm for every 24 square feet of surface area. Local codes include specific guidelines for determining roof and driveway runoff rates.

To calculate the average amount of rain in the area, consult a table or map of local area rainfall. Tables and maps are printed in model and local codes, civil engineering handbooks, and landscape architecture manuals. You can also get information from the U.S. Department of Agriculture or the National Weather Service of the National Oceanic and Atmospheric Administration (NOAA). The National Weather Service issues 100-year 1-hour rainfall maps for the United States. These maps show the average recorded rainfall in 1 hour for a given area over the previous century. See *Appendix A* for 100-year 1-hour rainfall maps for different parts of the country.

Design the removal system to discharge runoff into the building storm sewer. Local code may also allow the waste to discharge into a street gutter or

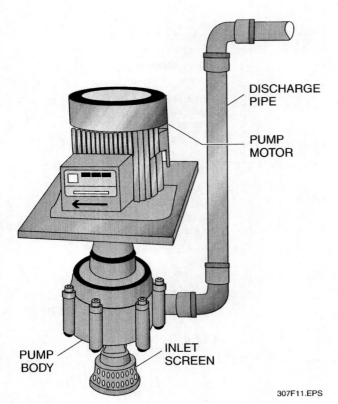

Figure 11 ◆ CPVC pump.

curb drain. Some older installations connect storm drains with sewage waste lines. Many codes now prohibit this type of connection. Many cities and counties maintain **retention ponds** and **retention tanks** for storm water drainage. They are used for flood control. Retention ponds are large man-made basins where sediment and contaminants settle out of the wastewater before the water returns to rivers, streams, and lakes. Retention tanks are similar to retention ponds, except they are enclosed. Install backwater valves on the drainage line to prevent contamination of the sump in case a retention pond or tank overflows.

3.3.0 Controls

Pump control is usually handled by a float sensor and/or a switch that activates the pump as the water level in the basin rises. Set the float to activate the pump at the highest water level possible inside the sump. This will prevent excessive cycling of the pump. Consider installing a standby pump or backup power system for emergency situations. An alarm will provide occupants with a warning signal during an emergency condition like a pump failure. Review the installation requirements and construction drawings to see if these measures are appropriate.

WARNING!
Review manufacturer's instructions before attempting to repair or replace a pump. Ensure that valves are closed and that electrical connections to the pump motor and controls have been turned off and tagged out. Wear appropriate personal protective equipment. Do not work in a walk-in wastewater sump without prior training in how to work safely in confined spaces.

4.0.0 ◆ TROUBLESHOOTING AND REPAIRING SEWAGE AND STORM WATER REMOVAL SYSTEMS

Occasionally, pumps break down. When they do, they must be repaired or replaced. A malfunctioning pump usually means an entire drainage system must be shut down until the pump can be repaired. Quick and efficient troubleshooting and repair or replacement is essential. Plumbers can perform most pump repairs. Pump problems can usually be traced to one or both of the following causes:

- Electrical, involving the pump motor and wiring
- Mechanical, involving the pump's moving parts

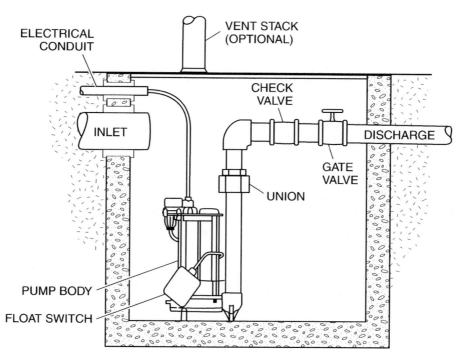

Figure 12 ◆ Check and gate valves in a centrifugal pump installation.

The most common types of problems and their solutions are discussed in the following sections.

4.1.0 Troubleshooting Electrical Problems

Most problems with sewage and sump pumps can be traced to failure of the pump's electric motor. Power failures often happen during storms, when sump pumps are needed most. The loss of electrical power causes the pump to stop working, which puts the structure at risk of flooding. If the water level rises to the level of the pump motor, remove the electrical cord from the power source. If the motor is in danger of getting wet, it may be necessary to remove it from the pump. If a motor gets wet, ensure that it is thoroughly dry before connecting it to a power source. Consider installing a battery-operated pump to operate the removal system during power outages.

If a pump is not running, check the power source to see if the voltage level is the same as the pump's required voltage. This is specified on the pump's nameplate. The problem may be a blown fuse or a broken or loose electrical connection. Be sure to check all fuses, circuit breakers, and electrical connections for breaks. If the power cord insulation is damaged, replace the cord with a new one of the same rating. If the power cord is open or grounded, you will need to contact an electrician to check the resistance between the cord's hot and neutral leads.

Occasionally, the impeller will lock. Have an electrician check the amps drawn by the pump motor. If the motor is drawing more amps than permitted by the manufacturer, the impeller may be blocked, the bearings may be frozen, or the impeller shaft may be bent. Remove the pump for inspection and repair or replace the damaged part. If a motor's overload protection has tripped, contact an electrician to inspect the motor.

WARNING!
Electrical pump motors that have been exposed to water may cause electrical shock when run. Wear appropriate personal protective equipment, and ensure that electric motors are thoroughly dry before testing.

Pumps sometimes fail to deliver their rated capacity because of electrical problems. If the impeller appears to be operating below normal speed, check the voltage supply against the motor's required voltage. Sometimes backflow can cause the impeller to rotate in the wrong direction. For single-phase electrical motors, simply shut off the power and allow the impeller to stop rotating. Then turn the pump on again; the problem should correct itself. For three-phase electrical motors, contact an electrician to inspect the motor (see *Figure 13*).

4.2.0 Troubleshooting Mechanical Problems

Nonelectrical, mechanical problems with pumps include worn, damaged, or broken controls or pump components. Float sensors are a common culprit in mechanical failures. Carefully inspect the floats for corrosion and sediment buildup, which can cause floats to stick. To test if a float switch is broken, bypass the switch. If the pump operates, the switch will need to be replaced. Ensure the float is properly adjusted and that it has not slipped from its desired location. In any case, when repairing a waste removal system, always recalibrate and test the floats for proper operation.

Pneumatic ejectors use a rubber diaphragm exhaust valve to allow compressed air to escape and wastewater to resume filling the sump. If an ejector will not shut off, inspect the diaphragm and its switch. If the switch is broken or the rubber diaphragm is weak, replace the part immediately. Clean the areas around the diaphragm to remove any sediment lodged between the retainer ring and the diaphragm.

Plugged vent pipes can also prevent a pump from shutting off. Inspect the vent pipe and clear it out if necessary. The pump may have an **air lock**, which means that air trapped in the pump body is interfering with the flow of wastewater. Turn off the pump for about one minute, then restart it. Repeat this several times until the air is cleared from the pump. If the removal system has a check valve, ensure that a ³⁄₁₆-inch hole has been drilled in the discharge pipe about 2 inches above the connection. The wastewater inflow may match the pump's capacity. If this is the case, recalculate the amount of wastewater entering the system and install a suitable pump.

WARNING!
Storm water runoff may contain corrosive chemicals such as oil and grease. Wear appropriate personal protective equipment when performing maintenance on storm water sumps.

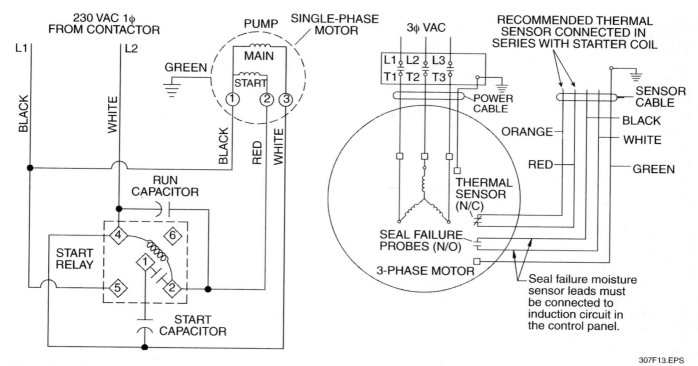

Figure 13 ◆ Single-phase and three-phase pump motors.

If the pump runs but won't discharge wastewater, the check valves may have been installed backward. Inspect the flow-indicating arrow on the check valves. If the check valve is still suspect, remove it and test to make sure it is not stuck or plugged. Check the pump's rating to ensure the lift is not too high for the pump. Inspect the inlet to the impeller and remove any solids that may be blocking it. Turn the pump on and off several times to remove an air lock.

When a pump is not delivering its rated capacity, check to see if the impeller is rotating in the wrong direction. Wastewater can drain back into a sump from a long discharge pipe if a check valve is not installed. Reverse flow will cause the impeller to rotate backward. Turn the pump off until the impeller stops rotating. Install a check valve on the discharge line to prevent the problem from recurring. Inspect the impeller for wear due to abrasives or corrosion, and replace a worn impeller immediately. Then turn the pump back on. Finally, check the pump rating to ensure the lift is not too high for the pump.

If the pump cycles continually, the problem could again be wastewater draining back into the basin from the discharge line. Install a check valve in the discharge line. If a check valve is already installed, inspect it for leakage. Repair or replace leaking valves. The sump may be too small for the wastewater inflow. Recalculate the inflow and, if necessary, install a larger basin.

Odors associated with a sewage removal system may mean that the sump lid is not properly sealed. Check the cover to see if its seal is intact. Inspect the vent to ensure the sump has open-air access. Some commercial structures, such as laundromats, use special lint traps. If the sewer lines are backed up, check to make sure all the lint traps are clean.

4.3.0 Replacing Sewage and Storm Water Pumps

If a pump cannot be repaired, it will have to be replaced. When replacing a sewage or storm water pump, always remember to install a properly sized replacement. If a pump repeatedly breaks down, it may be the wrong size or type. Review the construction drawings, plumbing installation design, and removal system specifications. If necessary, have an engineer appraise the design and installation. These steps will ensure that the pump is matched with the job it is supposed to perform.

 WARNING!
Review OSHA confined-space safety protocols before working in a walk-in sump. Wear appropriate personal protective equipment.

ON THE LEVEL
Plumbing Code Requirements for Sewage and Storm Water Removal Systems

Local plumbing codes govern the installation and operation of sewage and storm water removal systems. Many local codes are based on one of the model plumbing codes. Always refer to your local code before installing sewage and storm water removal systems. Here are some general guidelines from two model codes: the *International Plumbing Code® (IPC)* and the *Uniform Plumbing Code™ (UPC)*.

The *UPC* has two requirements for sewage removal systems: (1) the pump or discharge pipes connected to a water closet must be at least 2 inches in diameter; and (2) lines must have accessible backwater or swing check valves and gate valves.

The *IPC's* requirements for sewage removal systems are more specific. Lift stations must be covered with a gas-proof lid and must be vented. Dimensions of the basin must be at least 18 inches in diameter and 24 inches deep. Effluent (the system's outlet) must be kept at least 2 inches below the gravity drain system's inlet. For commercial installations, ejectors require a gate valve on the discharge side of the check valve before the gravity drain. Gate and check valves must be located above the lift station covers.

Additionally, the *IPC* mandates minimum ejector capacities: for discharge pipes 2 inches in diameter, the capacity must be 21 gpm; for discharge pipes 2½ inches in diameter, the capacity must by 30 gpm; and for discharge pipes 3 inches in diameter, the capacity must be 46 gpm.

According to the *UPC*, storm water removal systems must have sumps that are watertight and gravity-fed. The sumps can be made of steel, concrete, or other approved materials. Sumps must be vented, and the vent may be combined with other vent piping. In case of overload or mechanical failure, sumps located in "public use" areas must have dual pumps or ejectors that can function independently. Drain lines that connect to a horizontal gravity-fed discharge line must connect from the top through a wye. Gate valves are required on the discharge side of the backwater or check valve.

The *IPC's* storm water removal systems call for sub-drains that must discharge into a sump serviced by an appropriately sized pump. Sub-soil sumps do not require gastight covers or vents. Sumps must be at least 18 inches in diameter. Electrical service outlets must conform to National Fire Protection Association (NFPA) standards. Discharge pipes require a gate valve and full-flow check valve.

Many pumps are water-lubricated. This means that they must be immersed in water before they can be tested or run. Pumps are often self-priming. This means they automatically fill with water before operating. Failure to run the pump in water could burn out the pump's bearings quickly. Follow manufacturer's specifications closely when installing and testing sump and sewage pumps.

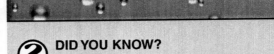

DID YOU KNOW?
The First Modern Sanitary Engineer

New York engineer Julius W. Adams laid the foundations for Brooklyn's sewer system in 1857. He ended up inventing a system that defined modern sanitary engineering. His design provided sewage drains for all of Brooklyn's 20 square miles. Adams used technology already in use by other cities—reservoirs feeding into underground mains that branched off to connect commercial and residential buildings. However, Adams did something that no one had done before: he distilled his experience into a sewer design manual, which he then published. For the first time, cities throughout the country could consult a standard reference for sewer design.

Review Questions

Sections 3.0.0–4.0.0

1. When sizing a sump pump, ensure that it has a capacity of _____ times the maximum calculated flow.
   a. 1½
   b. 2
   c. 2½
   d. 3

2. The formula for determining the volume of a sump is _____.
   a. $V = (\pi r^2)h$
   b. $V = lwh$
   c. $V = \frac{1}{2}(lwh)$
   d. $V = \pi r^2$

3. When sizing a storm water sump for use in sandy soil, unless otherwise specified use a collection rate of _____ gpm for every _____ square feet.
   a. 1; 50
   b. 2; 50
   c. 1; 100
   d. 2; 100

4. Install float sensors to activate the pump at the _____ water level in the basin.
   a. lowest possible
   b. average
   c. highest possible
   d. initial

5. _____ can cause an impeller to rotate in the wrong direction.
   a. Air lock
   b. Backflow
   c. Weak seals
   d. Corrosion

5.0.0 ◆ THE WORKSHEET

Refer to the appropriate sections in the module to answer the following questions.

1. What are the two types of pump that can be used in sewage removal systems?

2. How does an impeller work?

3. What are the reasons for installing a duplex pump?

4. What three considerations should a plumber keep in mind when calculating drainage fixture units?

5. Refer to *Appendix B*. Identify the various parts of a sewage waste removal system pointed out in the illustration.

6. List four different types of pump switches.

7. What collection rate can plumbers use to estimate runoff from roofs and driveways?

8. Name at least two causes of impeller lock.

9. Describe the procedure for inspecting a suspect float.

10. What are the mechanical reasons a pump will cycle constantly?

Summary

Sewage and storm water removal systems are important components of plumbing installation. They pump wastewater in sub-drains into the sewer line, where they can discharge into a public sewer, retention pond, or retention tank. Pneumatic ejectors and centrifugal pumps are used to lift wastewater from basins into the discharge line. Sumps hold the wastewater until the pump discharges it. Switches installed on the inside of the basin trigger the pump when the wastewater reaches a preselected height.

Plumbers are responsible for maintaining, repairing, and replacing pumps. Electrical and mechanical problems can be diagnosed by inspecting various components. Always remember to take appropriate precautions when working with electrical pumps near water. Consult with engineers and electricians if the problem requires more than the repair or replacement of a part or a whole pump. To protect yourself against harmful solids and liquids in wastewater, always wear appropriate personal protective equipment when working with sewage and sump pumps. Because sewage and storm water removal systems help maintain public health and safety, respond to pump problems quickly and professionally.

Notes

Trade Terms Introduced in This Module

Air lock: A pump malfunction caused by air trapped in the pump body, interrupting wastewater flow.

Centrifugal pump: A device that uses an impeller powered by an electric motor to draw wastewater out of a sump and discharge it into a drain line.

Drainage fixture unit: A measure of the discharge in gallons per minute of a fixture, divided by the number of gallons in a cubic foot.

Duplex pump: Two centrifugal pumps and their related equipment installed in parallel in a sump.

Ejector: A common alternative name for a sewage pump.

Float switch: A container filled with a gas or liquid that measures wastewater height in a sump and activates the pump when the float reaches a specified height.

Impeller: A set of vanes or blades that draws wastewater from a sump and ejects it into the discharge line.

Lift station: A popular name for a sewage removal system. The term is also used to refer to a pumping station on a main sewer line.

Mercury float switch: A float switch filled with mercury. It activates the pump when it reaches a specified height in a basin.

Pneumatic ejector: A pump that uses compressed air to force wastewater from a sump into the drain line.

Pressure float switch: An adjustable float switch that can either be allowed to rise with wastewater or be clipped to the sump wall.

Probe switch: An electrical switch used in pneumatic ejectors that closes on contact with wastewater.

Retention pond: A large man-made basin into which storm water runoff drains until sediment and contaminants have settled out.

Retention tank: An enclosed container that stores storm water runoff until sediment and contaminants have settled out.

Reverse flow pump: A duplex pump designed to discharge solid waste without allowing it to come into contact with the pump's impellers.

Sewage pump: A device that draws wastewater from a sump and pumps it to the sewer line. It may be either a centrifugal pump or a pneumatic ejector.

Sewage removal system: An installation consisting of a sump, pump, and related controls that stores sewage wastewater from sub-drains and lifts it into the main drain line.

Storm water removal system: An installation consisting of a sump, pump, and related controls that stores storm water runoff from a sub-drain and lifts it into the main drain line.

Sub-drain: A building drain located below the sewer line.

Sump: A container that collects and stores wastewater from a sub-drain until it can be pumped into the main drain line.

Sump pump: A common name for a pump used in a storm water removal system.

Appendix A

100-Year, 1-Hour Rainfall in Inches

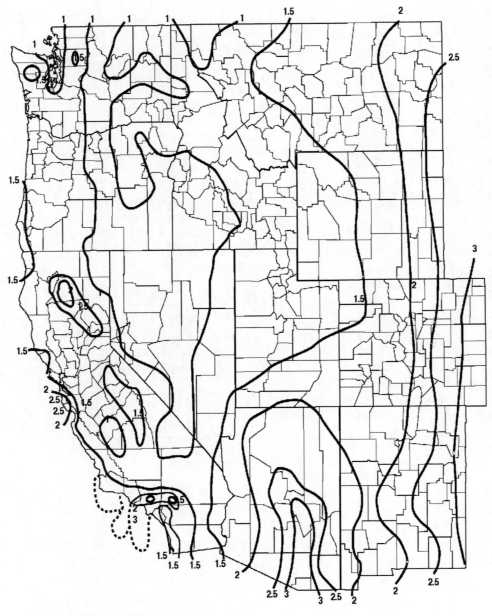

WESTERN UNITED STATES

FOR SI: 1 INCH = 25.4 MM
SOURCE: NATIONAL WEATHER SERVICE, NATIONAL OCEANIC AND ATMOSPHERIC ADMINISTRATION, WASHINGTON, DC

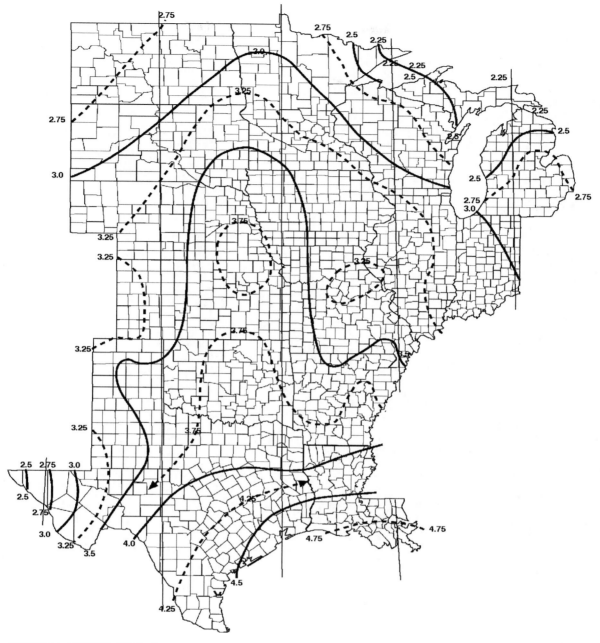

CENTRAL UNITED STATES

FOR SI: 1 INCH = 25.4 MM
SOURCE: NATIONAL WEATHER SERVICE, NATIONAL OCEANIC AND ATMOSPHERIC ADMINISTRATION, WASHINGTON, DC

EASTERN UNITED STATES

FOR SI: 1 INCH = 25.4 MM
SOURCE: NATIONAL WEATHER SERVICE, NATIONAL OCEANIC AND ATMOSPHERIC ADMINISTRATION, WASHINGTON, DC

307A03.EPS

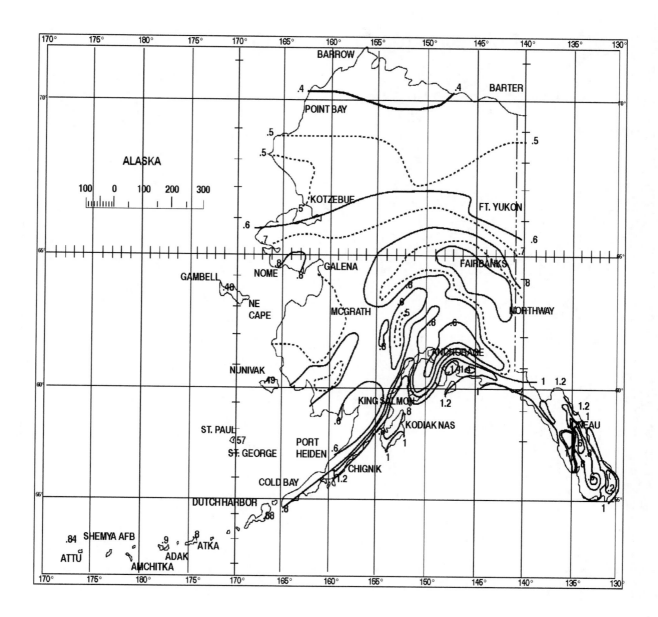

ALASKA

FOR SI: 1 INCH = 25.4 MM
SOURCE: NATIONAL WEATHER SERVICE, NATIONAL OCEANIC AND ATMOSPHERIC ADMINISTRATION, WASHINGTON, DC

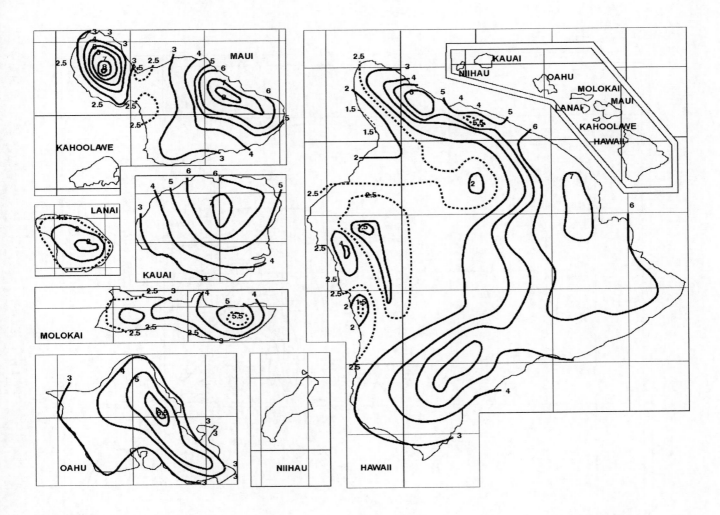

HAWAII

FOR SI: 1 INCH = 25.4 MM
SOURCE: NATIONAL WEATHER SERVICE, NATIONAL OCEANIC AND ATMOSPHERIC ADMINISTRATION, WASHINGTON, DC

Appendix B

Worksheet Illustration

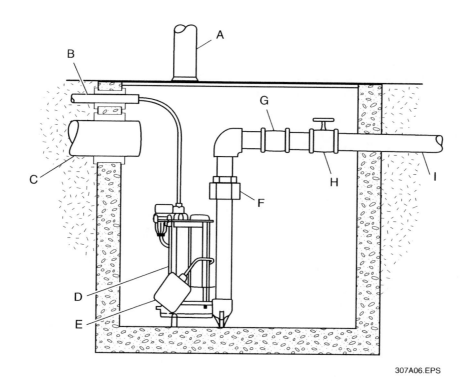

Additional Resources

This module is intended to be a thorough resource for task training. The following reference works are suggested for further study. These are optional materials for continued education rather than for task training.

Modern Plumbing. 1997. Howard C. Massey. Carlsbad, CA: Craftsman Book Company.

2003 International Plumbing Code. 2003. Falls Church, VA: International Code Council.

Building Services Engineering: A Review of Its Development. 1982. Neville S. Billington. New York: Pergamon Press.

References

Dictionary of Architecture and Construction, Third Edition. 2000. Cyril M. Harris, ed. New York: McGraw-Hill.

Pipefitters Handbook. 1967. Forrest R. Lindsey. New York: Industrial Press Inc.

Figure Credits

International Code Council, Inc., *2003 International Plumbing Code*, 307A01–307A05, Tables 1, and 2

Liberty Pumps, 307F12

Stevens Pump Company, 307F01, 307F10

Zoeller Pump Company, 307F06, 307F07, 307F13

CONTREN® LEARNING SERIES — USER UPDATE

The NCCER makes every effort to keep these textbooks up-to-date and free of technical errors. We appreciate your help in this process. If you have an idea for improving this textbook, or if you find an error, a typographical mistake, or an inaccuracy in NCCER's Contren® textbooks, please write us, using this form or a photocopy. Be sure to include the exact module number, page number, a detailed description, and the correction, if applicable. Your input will be brought to the attention of the Technical Review Committee. Thank you for your assistance.

Instructors – If you found that additional materials were necessary in order to teach this module effectively, please let us know so that we may include them in the Equipment/Materials list in the Annotated Instructor's Guide.

Write: Product Development and Revision
National Center for Construction Education and Research
P.O. Box 141104, Gainesville, FL 32614-1104

Fax: 352-334-0932

E-mail: curriculum@nccer.org

Craft _____ Module Name _____

Copyright Date _____ Module Number _____ Page Number(s) _____

Description
_____
_____
_____
_____

(Optional) Correction
_____
_____
_____

(Optional) Your Name and Address
_____
_____
_____

Plumbing Level Three

02308-06

Corrosive-Resistant Waste Piping

02308-06
Corrosive-Resistant Waste Piping

Topics to be presented in this module include:

1.0.0	Introduction	8.1
2.0.0	Types of Corrosive Wastes	8.1
3.0.0	Pipe Materials for Corrosive Wastes	8.3
4.0.0	Installing Corrosive-Resistant Waste Piping Systems	8.7
5.0.0	Hazard Communication	8.14

Overview

Corrosive wastes must be handled differently from other types of waste. They contain harsh chemicals that can damage the inside of ordinary drainage and vent pipes, and they can cause illness, injury, and death if people are exposed to them. Unlike household chemical wastes such as bleaches and detergents, which are not harsh enough to damage drainage pipes, corrosive wastes created by industrial, commercial, and laboratory processes require a separate removal system.

Plumbers are responsible for installing waste piping that safely discharges corrosive wastes. The codes and standards governing these systems are very strict. For each system installation, plumbers must select pipe and fittings made of special materials that can resist corrosive wastes, and that are suited for the type of waste being created. The most common materials used to make such pipe and fittings are borosilicate glass, thermoplastic and thermoset plastic, silicon cast iron, and stainless steel. Each of these materials has unique corrosive-resistant properties and is appropriate for different types of corrosive wastes.

While installation techniques for corrosive-resistant waste systems are similar to those for other types of waste systems, there are some important additional requirements. Plumbers must ensure that the system is completely separate from other drainage systems in the building. They must also install acid dilution and neutralization sumps to treat the corrosive wastes, and test the completed system in accordance with local codes. To protect workers and the public, Hazard Communication (HazCom) plans are required by law for companies that handle corrosive and toxic materials.

Focus Statement
The goal of the plumber is to protect the health, safety, and comfort of the nation job by job.

Code Note
Codes vary among jurisdictions. Because of the variations in code, consult the applicable code whenever regulations are in question. Referring to an incorrect set of codes can cause as much trouble as failing to reference codes altogether. Obtain, review, and familiarize yourself with your local adopted code.

Portions of this publication reproduce tables and figures from the *2003 International Plumbing Code* and the *2000 IPC Commentary*, International Code Council, Inc., Falls Church, Virginia. Reproduced with permission. All rights reserved.

Objectives

When you have completed this module, you will be able to do the following:

1. Discuss corrosive wastes and explain where they are found.
2. Discuss common types of materials used for corrosive-resistant waste piping.
3. Explain the methods of joining corrosive-resistant waste piping.
4. Discuss safety issues and hazard communications.

Trade Terms

Adhesive
Austenitic stainless steel
Beaded end
Biohazard
Borosilicate glass
Corrosive waste
Coupling
Duplex stainless steel
Ferritic stainless steel
Furan
Fusion coil
Hazard communication
Martensitic stainless steel
NFPA diamond
Thermoplastic
Thermoset
Uniform Laboratory Hazard Signage

Required Trainee Materials

1. Appropriate personal protective equipment
2. Sharpened pencils and paper
3. Copy of local applicable code
4. Calculator

Prerequisites

Before you begin this module, it is recommended that you successfully complete *Core Curriculum; Plumbing Level One; Plumbing Level Two; Plumbing Level Three*, Modules 02301-06 through 02307-06.

This course map shows all of the modules in the third level of the *Plumbing* curriculum. The suggested training order begins at the bottom and proceeds up. Skill levels increase as you advance on the course map. The local Training Program Sponsor may adjust the training order.

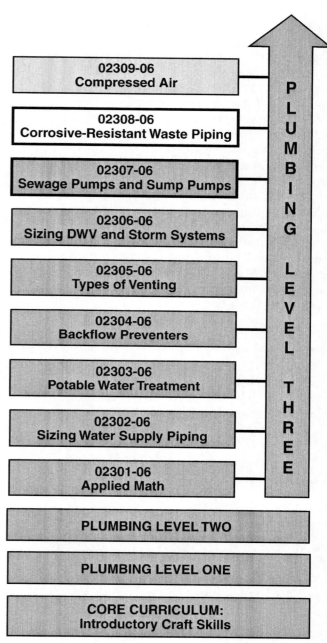

1.0.0 ♦ INTRODUCTION

Plumbers install drain, waste, and vent (DWV) systems to safely handle a wide variety of wastes. You have learned about DWV systems that handle sewage and storm water. You may also install systems designed to handle **corrosive wastes**. Corrosive wastes contain harsh chemicals that can damage the inside of ordinary drainage and vent pipes. Some corrosive wastes contain residue that could restrict the flow in the drainage system. Corrosive wastes can also damage sewage treatment systems. They can also cause illness, injury, and death if people are exposed to them.

Corrosive wastes must be handled differently from other types of waste. They require the installation of a separate drainage system (see *Figure 1*). Corrosive wastes must also be treated before they can be discharged. Drainage pipes must be made of special materials that can resist corrosive wastes. In this module, you will learn about the different types of corrosive-resistant pipe and how to install them. You will also learn how to identify and label dangerous chemicals in the workplace. Study each section carefully. The rules and guidelines are there to protect you and your customers from harm.

NOTE

Plumbers use the term corrosive-resistant waste piping to refer to pipe that is designed specifically to handle corrosive wastes. The term does not imply that such pipe is designed to resist other types of corrosion—for example, electrolysis, galvanic corrosion, scale, or rust.

DID YOU KNOW?
Corrosive Chemicals

Corrosive chemicals did not appear in large quantities until the nineteenth century. As a result of the Industrial Revolution, European and American companies began to use larger and larger amounts of chemicals. People soon noticed that chemicals were damaging waste pipes when they were discarded. Engineers began to develop new metal alloys that could resist these harsh wastes, marking the beginning of the modern corrosive-resistant waste piping industry.

2.0.0 ♦ TYPES OF CORROSIVE WASTES

The United States Environmental Protection Agency (EPA) defines corrosive wastes as liquids that have a pH value of 2 or lower (extremely acidic) or 12 or higher (extremely alkaline). Corrosive wastes can eat through steel at a rate of ¼ inch per year. They can erode the insides of many types of metal and plastic waste pipes. They can create toxic or noxious fumes that can spread throughout a building. In other words, corrosive wastes are hazardous not only to sanitation but to health.

Many common corrosive wastes are created in the home. Even though some of them could damage a waste system, they usually are not treated before they are discharged. Most wastes created in the home are less harsh than corrosive wastes created by industry. As a result, standard DWV piping can handle them. The following common home products create corrosive wastes:

- Scalding water
- Bleaches
- Detergents
- Drain cleaners
- Scouring powders
- Other household chemicals

Almost any acid-resistant material can be used for DWV piping in the home. Concrete, cast iron, plastic, steel, copper, and clay can all be used. Some home septic systems are equipped with a separate tank to handle laundry wastes. Otherwise, little extra protection is needed from household corrosive wastes.

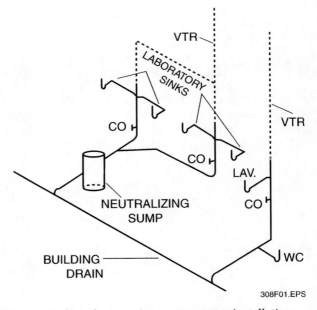

Figure 1 ♦ Sample corrosive-waste system installation.

This module focuses on corrosive wastes created by industrial, commercial, and laboratory processes. These wastes are harsher than household chemical wastes. They must be treated before they can be discharged into the drainage system, so a separate corrosive waste system must be installed. Many types of industrial or manufacturing businesses create corrosive wastes:

- Auto painting and detailing shops
- Construction companies
- Laboratories
- Metal manufacturers
- Paint manufacturers
- Pharmaceutical manufacturers
- Photographic processing facilities
- Pulp and paper mills
- Vehicle maintenance facilities

Corrosive wastes that you are likely to find on the job include rust removers, cleaning fluids, solvents, battery acid, degreasers, paint wastes, printing ink, pesticides, and agricultural chemicals. Plumbers are responsible for installing waste piping that safely discharges corrosive wastes.

3.0.0 ◆ PIPE MATERIALS FOR CORROSIVE WASTES

Each waste system requires special pipe and fittings that are suited for the type of waste being created. The codes and standards that govern corrosive waste systems are very strict. Consult your local code to find out what pipe materials may be used in your area. Many manufacturers provide tables that list the chemicals that their pipes can handle safely. Corrosive-resistant pipe and fittings are more expensive than regular waste pipe and fittings.

The following are the most common materials used to make pipe and fittings for corrosive waste systems:

- **Borosilicate glass**
- **Thermoplastic** and **thermoset** plastics
- Silicon cast iron
- Stainless steel

Each of these materials has unique corrosive-resistant properties. They are suitable for a wide variety of waste systems. The properties and uses of each type of pipe material are discussed in more detail in the following sections. Review each section carefully. Talk with an experienced plumber to learn more about the different types of pipe. Remember that the goal is to select the pipe that will provide years of safe and trouble-free operation.

3.1.0 Borosilicate Glass Pipe

Many corrosive wastes are discharged at extreme temperatures. Use borosilicate glass pipe (see *Figure 2*) for these wastes. Borosilicate glass pipe is made from a specially reinforced transparent glass. Changes in temperature cause very little expansion or contraction. As a result, borosilicate glass pipes can safely handle liquids ranging from freezing to 212°F. However, sudden extreme changes in temperature may cause the pipe to crack.

Borosilicate glass pipes come with either plain or **beaded ends**. A beaded end is a lip around the pipe rim. The pipe and fittings in a glass drainage system can be reused. Use a glass pipe cutter (see *Figure 3*) to cut the pipes to new lengths as needed.

Use **couplings** to join pipes and fittings. Couplings are removable clamps with plastic seals on the inner rim (see *Figure 4*). Couplings provide leak-free joints even when the pipe is slightly deflected. The seals on the inside of the coupling are made from tetrafluoroethylene (TFE) plastic. TFE resists chemicals and acids better than any other type of plastic. The outer shell of the coupling is made of stainless steel. Couplings join beaded end to beaded end, beaded end to plain end, and metal or plastic to plain end (see *Figure 5*).

3.2.0 Plastic Pipe

Codes allow plastic pipe to be used for corrosive waste drainage. Plastic pipe has the following advantages over other pipe materials:

- Resistance to chemical and galvanic corrosion
- Low friction loss
- High handling temperatures
- Flexibility
- Inflammability

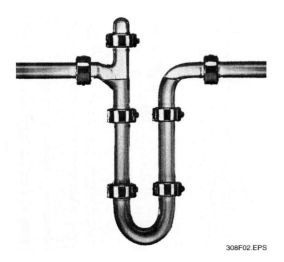

Figure 2 ◆ Borosilicate glass pipe and fittings.

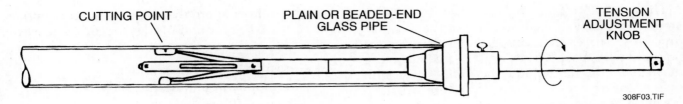

Figure 3 ◆ Glass pipe cutter.

Figure 4 ◆ Coupling used to join glass pipe.

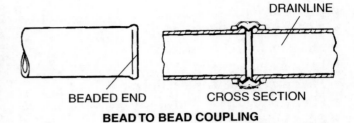

BEAD TO BEAD COUPLING

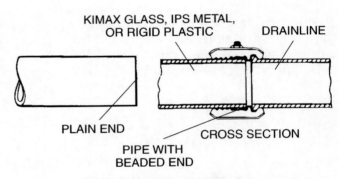

BEAD TO PLAIN COUPLING

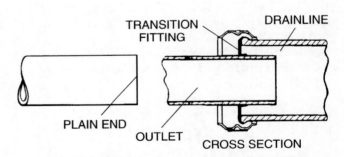

METAL OR PLASTIC TO PLAIN COUPLING

Figure 5 ◆ Three types of couplings.

Refer to the manufacturer's catalog when selecting plastic pipe. The catalog will list the properties of each type of pipe. It will help you select the pipe material that is best suited for the type of corrosive waste in a given system.

Thermoplastic and thermoset plastic pipe can be used to drain corrosive wastes. Both types of plastic pipe are composed of a resin base with chemical additives. The difference is that thermoplastic pipe can be softened and reshaped when exposed to heat. Thermoset pipe, on the other hand, is hardened into shape when manufactured. Suitable thermoplastic pipe materials include the following:

- Polyvinyl chloride (PVC)
- Chlorinated polyvinyl chloride (CPVC)
- Polyvinylidene fluoride (PVDF)
- Polypropylene (PP)

You are already familiar with PVC and CPVC pipe. PVDF is more resistant to corrosion and abrasion than other types of plastic pipe. PP is a light, chemically inert plastic. PP is safe for use in kitchens and other food preparation areas. Thermoplastic pipe can handle acetic acids and chemical compounds containing aluminum, ammonia, barium, calcium, sodium, and mild sulfuric acids. Resistance varies depending on the type of pipe.

 WARNING!
Corrosive chemicals can cause injury or death if they are not handled properly. Read all warning labels before you handle any chemical. Wear appropriate personal protective equipment when you work with chemicals, and always follow the safety guidelines.

Thermoset pipes are made from epoxies, polyesters, vinyl esters, or **furans**. Furans are resins that are highly resistant to solvents. The pipes are wrapped in fiberglass or graphite fibers for reinforcement. The fiber wrapping makes the pipes rigid and strong. Thermoset plastic pipes are highly resistant to corrosion. Because they are more rigid than other types of pipe, hangers must be closely spaced to provide adequate support.

Thermoset pipe can handle many of the same corrosive chemicals as other types of plastic pipe. Properties of specific types of pipe will vary depending on the manufacturer. Always refer to the specifications for each pipe before you select it. Refer to your local code to determine whether the pipe is suitable for the type of installation.

Some pipes are not suitable for certain types of corrosive wastes. Pipes made out of furan resins, for example, should not be used with oxidized wastes. PVDF is unsuitable for certain strong alkalis or acids.

3.3.0 Silicon Cast-Iron Pipe

Silicon cast-iron pipe (see *Figure 6*) can be used to drain almost all types of corrosive waste. It looks like other types of cast-iron pipe, but the iron alloy used in the pipe is 14 percent silicon and 1 percent carbon. The silicon and carbon make the pipes and fittings highly resistant to corrosives.

Silicon cast-iron pipe can handle most corrosive wastes regardless of concentration or temperature. It is often used to drain sulfuric and nitric acids. Waste systems made from silicon cast-iron pipe usually last the life of the building. Silicon cast-iron pipe is often referred to by its trade name, Flowserve. Silicon cast iron is available in hub-and-spigot (see *Figure 7*) and no-hub styles. Use mechanical joints to connect no-hub pipe. A typical mechanical joint has an inner sleeve made of Teflon® surrounded by neoprene. Teflon® has very high corrosion resistance, and the outer neoprene sleeve adds strength. Neoprene is resistant to heat, cold, weathering, and abrasion. Surrounding the inner and outer sleeves is a stainless steel coupling fastened by two bolts. The coupling allows the pipe to have a small amount of deflection.

Use mechanical or hydraulic spring-loaded pipe cutters to cut silicon cast-iron pipe; cold chisels may also be used. Support the pipe on a sandbag to absorb the shock. A properly cut silicon cast-iron pipe has a smooth edge. If the edge is smooth, the pipe can be installed with no further preparation.

Do not use silicon cast-iron pipe in a pressurized corrosive-resistant waste system. Your local code will specify the type of pipe to be used in a pressurized system. Silicon cast-iron pipe can be painted.

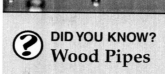

DID YOU KNOW?
Wood Pipes

Some heavily wooded countries still use wood for supply and waste pipes. Wood is highly resistant to the harsh effects of many chemicals and can handle some chemicals better than metal pipe.

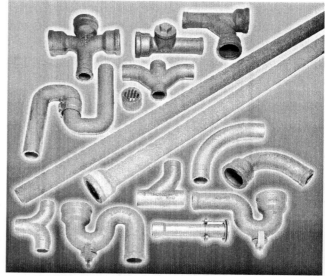

Figure 6 ♦ Silicon cast-iron pipe and fittings.

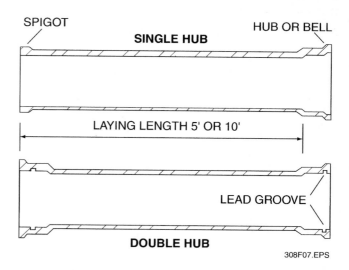

Figure 7 ♦ Hub-and-spigot silicon cast-iron pipe.

CAUTION

Remember that pipe must be labeled with the name of the manufacturer, the trademark of the Cast Iron Soil Pipe Institute, and the pipe diameter. Most codes will not allow you to install unlabeled pipe. Refer to your local code.

3.4.0 Stainless Steel Pipe

Plumbers can choose from four types of stainless steel pipe. Each type is made of a different metal alloy that offers its own special corrosion-resistance capabilities. The following are the four types of stainless steel alloys available for corrosive-resistant waste piping:

- Ferritic
- Austenitic
- Martensitic
- Duplex

Ferritic stainless steel is very resistant to rust and has an extremely low carbon content. Use ferritic steel for chloride and organic acid wastes. Ferritic stainless steel is often used in food processing facilities. **Austenitic stainless steel** contains nickel, which makes the metal more stable than ferritic steel. It stays strong at very low temperatures. Austenitic steel is the most popular type of steel for corrosive waste pipes. **Martensitic stainless steel** is very strong, but it does not resist corrosion as well as austenitic steel. It is mostly used to handle mild corrosive wastes, such as organic matter. **Duplex stainless steel** is a mix of austenitic and ferritic steels. It is stronger than austenitic steel, which means that thinner, lighter pipe can be used. Duplex stainless steel offers very good corrosion resistance.

Review Questions

Sections 1.0.0–3.0.0

1. Corrosive wastes are liquids with a pH value of either _____ or lower, or _____ or higher.
   a. 2; 12
   b. 2; 14
   c. 3; 10
   d. 5; 10

2. Borosilicate glass pipes come with either _____ ends or beaded ends.
   a. plain
   b. threaded
   c. flanged
   d. beveled

3. PVC, CPVC, PVDF, and PP are examples of _____ pipe.
   a. borosilicate glass
   b. thermoset plastic
   c. thermoplastic
   d. silicon cast-iron

4. To connect no-hub silicon cast-iron pipe, use a _____.
   a. solvent cement
   b. mechanical joint
   c. union flange
   d. threaded joint

5. The most popular type of steel for corrosive waste pipes is _____ steel, which contains nickel and stays strong at low temperatures.
   a. Ferritic
   b. Austenitic
   c. Martensitic
   d. Duplex

4.0.0 ♦ INSTALLING CORROSIVE-RESISTANT WASTE PIPING SYSTEMS

You can install corrosive-resistant waste systems using techniques that you have already learned, but some of the specific requirements are different from those for other waste systems. Ensure that the system is completely separate from other drainage systems in the building. Install a trap on every fixture in the system. The wastes must be diluted or neutralized before they can be discharged. Refer to your local code for installation requirements in your area.

Engineers design corrosive-resistant waste systems. Refer to the project plans for the materials to be used in the system. If you have any questions about the design, ask the engineer. Install the corrosive-resistant waste system so that it can be expanded or changed in the future. Remember that these installation instructions are for general reference only. Refer to your local code for specific guidelines. Test the system in accordance with local codes when the installation is complete.

4.1.0 Joining and Installing Borosilicate Glass Pipe

Glass pipes and fittings are usually delivered to the job site in rugged shipping cartons. To avoid damaging them, unpack them only when they are needed. As mentioned, borosilicate glass pipe comes with beaded or plain ends. Couplings join glass pipe by clamping the two ends together. Pipes can be joined beaded end to beaded end or beaded end to plain end. Couplings are also available to join glass pipe to plastic and metal pipe. Use the following steps to join pipes using a beaded end to beaded end coupling (see *Figure 8*).

Step 1 Dip the coupling in water or wet the inside with a damp cloth.

Step 2 Push and rotate the coupling to snap it over the pipe bead. Start with the side of the coupling that is opposite the bolt.

Step 3 When the coupling has been snapped onto the first pipe, insert the other section of pipe into the other side of the coupling.

Step 4 Tighten the coupling bolt using a torque wrench. Do not exceed the manufacturer's torque specifications when tightening the bolt.

The procedure for joining a beaded-end pipe to a plain-end glass, metal, or rigid plastic pipe is similar:

Step 1 Dip the coupling in water or wet the inside with a damp cloth.

Step 2 Snap the coupling over the beaded end.

Step 3 Insert the plain-end pipe into the opposite side of the coupling, ensuring that the plain end is fully seated. Do not force the pipe into the coupling.

Step 4 Tighten the coupling bolt using a torque wrench. Do not exceed the manufacturer's torque specifications when tightening the bolt.

Though glass pipe looks very different from other types of pipe, it is installed the same way. It can be installed in aboveground and underground systems. Remember that glass pipe can break if it is strained. Reduce strain by allowing vertical and horizontal installations to have some limited movement. After you install borosilicate glass pipe, test it according to your local code. Air tests should not exceed 5 psi, and water tests should not exceed 22 psi.

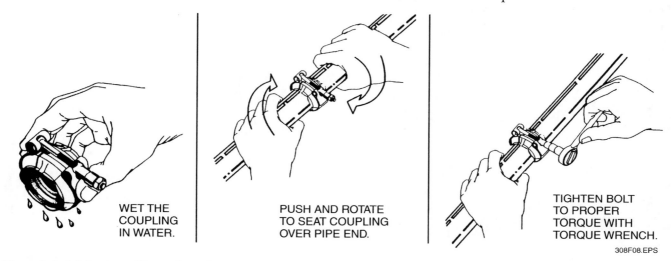

Figure 8 ♦ Joining borosilicate glass pipes.

4.1.1 Aboveground Installation

Glass piping is available in 5-foot and 10-foot lengths. Diameters range from 1½ to 6 inches. It can be installed to almost any length or height and can be installed behind walls. Glass pipe is safe to install in both horizontal and vertical runs. Refer to the installation's design drawings, and consult the engineer if you have any questions about the use of glass pipe in an installation.

Use padded hangers for horizontal runs of glass pipe (see *Figure 9*). The padding will help prevent the metal hangers from scratching the glass. Space hangers 8 to 10 feet apart. If the pipe has three or more couplings within an 8- to 10-foot span, install an extra hanger. Do not force glass pipe into a hanger. Always move the hanger to the glass pipe instead. Ensure that horizontal runs are free to move sideways. To change the pitch of a run of glass pipe, loosen the coupling bolt, tilt the pipe to the pitch required, and retighten the coupling bolt allow the pipe to be adjusted up to 4 degrees in any direction.

For horizontal runs of glass pipe behind interior walls, use pipe sleeves with a diameter that is at least 2 inches greater than the outside diameter of the pipe. Pack the space between the pipe and the sleeve with fiberglass or glass wool. For pipes that pass through exterior fire-, water-, or explosion-proof walls, pack the sleeve with appropriate caulking material. Space the couplings no more than 6 inches apart on either side of the wall.

Use riser clamps for vertical installation of glass pipe. The clamps must be lined with ¼ inch of neoprene rubber (see *Figure 10*). Refer to your local code for the proper number and spacing of riser clamps, and to the manufacturer's instructions for additional support requirements. Install riser clamps below couplings, not above them. Install a clamp below the bottom coupling in a glass pipe stack. Where possible, install clamps below couplings on every third floor. Do not support a vertical glass pipe stack with a horizontal pipe run. Locate the first hanger on a horizontal line 6 to 8 feet from where a vertical stack connects to it.

Fit pipe sleeves on all vertical stacks that pass through a floor or slab. The sleeve's diameter should be at least 2 inches greater than the diameter of the pipe. Pack the space between the pipe and sleeve with fiberglass or glass wool. Install a coupling within 6 inches of the floor or slab. The coupling allows the pipe to have some flexibility of motion. Some installations will require you to pass pipe vertically through fire-, water-, or explosion-proof floors. Pack the sleeve with appropriate caulking material or water plug cement. Follow your local code carefully when you install pipe through fireproof floors or walls.

Glass pipe can be used to vent a stack through the roof. Install vent flashing to prevent rainwater from running into the building through the hole cut for the vent. Depending on the location, job, and contract, this may be a job for the plumber, carpenter, or roofer. Check with your supervisor to find out if installing the vent flashing is part of your job. Wrap the vent pipe with tape or insulating material according to your local code and the building plans. Use either seamless lead roof flashing with a caulked counter-flashing sleeve or seamless lead or copper roof flashing.

To connect the glass pipe stack to the floor drain, install a glass-to-steel coupling to connect the glass pipe to a stainless steel pipe. Connect the metal pipe to the drain according to the manufacturer's instructions. Another option is to insert the glass pipe into the floor drain using the following steps:

Step 1 Cut the glass pipe to the desired length. Smooth the sharp outer edges of the pipe using the proper finishing tool.

Step 2 Insert the pipe into the outlet of the floor drain.

Step 3 Seal the glass pipe in place using asbestos rope and acid-proof cement. Allow the cement to cure for the time listed in the manufacturer's instructions. Be sure to use appropriate personal protective equipment when you are using asbestos and cement.

4.1.2 Underground Installation

Glass pipe and fittings can be used for underground drainage systems. Use glass pipe that is fitted with a protective casing. Expanded polystyrene is commonly used as a casing material. Most codes specify 5-foot lengths of pipe for underground installations. Wrap the fittings in thin polyvinyl film, scotch wrap, or other recom-

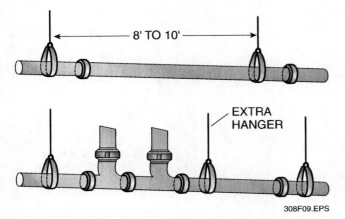

Figure 9 ♦ Installing a horizontal pipe run using padded hangers.

mended material. Remove the casing before you cut a pipe. Cut the casing 2 inches shorter than the new length of pipe. Replace the casing so that 1 inch of pipe is exposed at each end. The gap provides enough space to install a coupling.

To lay glass pipe underground, begin by excavating the pipe trench to a width of at least 24 inches at the bottom. If the soil is sandy, dig the trench 1 to 2 inches below the final grade. For rocky soil or clay, dig the trench 4 to 6 inches below the final grade (see *Figure 11*). Ensure that the trench bedding is firm enough to support the glass pipe uniformly along the full length of the run.

Assemble several glass pipe joints to form a section. Tighten the couplings firmly, and lower the section into the trench. Bury the pipe using thin layers of rock-free sand or soil until it is 12 inches above the glass pipe. Tamp the sand firmly by hand. Then backfill the trench with available soil using a backhoe or other mechanical means (refer to *Figure 11*).

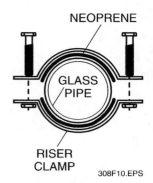

Figure 10 ◆ Installing a riser clamp on a waste pipe.

4.2.0 Joining and Installing Plastic Pipe

Thermoplastic and thermoset pipe can be joined and installed in several ways. Use the following to join plastic pipe:

- Fusion method
- Solvent cement
- Mechanical joints
- Pipe threading

The fusion method uses heat to join PVDF and PP pipes to a fitting (see *Figure 12A*). Insert a pipe end into a fusion fitting, which contains a **fusion coil** and electrical terminals. Clamp the fitting according to the manufacturer's instructions, and connect electrical leads to the coil. The fusion coil heats the pipe end and the fitting socket, causing them to fuse. Follow manufacturer's directions for heating, cooling, and testing the connection. Older systems require the worker to position the fusion coil itself. Newer systems embed heating elements in a coupling and fuse both ends of the connection at the same time (see *Figure 12B*). The fusion method is effective for pipe systems that handle corrosive wastes.

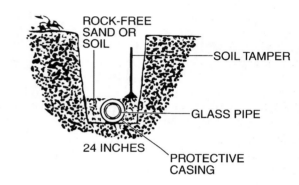

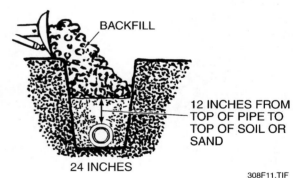

Figure 11 ◆ Laying borosilicate glass pipe underground.

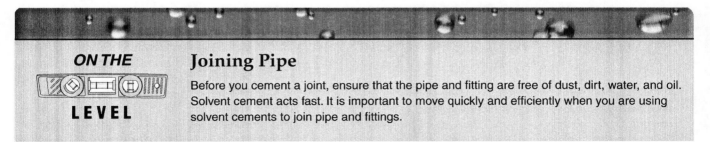

ON THE LEVEL

Joining Pipe

Before you cement a joint, ensure that the pipe and fitting are free of dust, dirt, water, and oil. Solvent cement acts fast. It is important to move quickly and efficiently when you are using solvent cements to join pipe and fittings.

HUBLESS FUSION WITH COIL FITTING AND CLAMPS

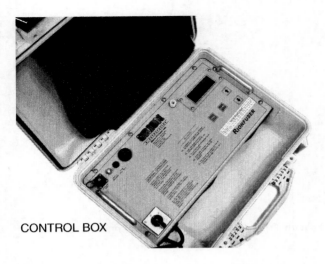

CONTROL BOX

Step 1. Measure the full depth of the fitting socket.

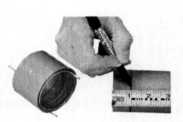

Step 2. Transfer the measurement to the pipe, marking around its circumference. (Be sure the pipe is cut square and has been deburred.)

Step 3. Sand the shiny surface from the area of the pipe to be fused.

Step 4. Insert pipe completely into the fitting end that contains the electrofusion coil. Check that it meets the measured mark around the pipe.

Step 5. Position the clamp band, following manufacturer's instructions. Tighten the clamp by hand. Pipe sizes of 3" or more require additional tightening with a clamping tool.

Step 6. Connect electrical leads to the coil terminals.

Step 7. Verify settings on the control box and press the "start" button on either the output lead or the control box. Follow manufacturer's instructions for cooling and testing.

308F12A.EPS

Figure 12 ♦ The fusion method. (1 of 2)

HUBLESS FUSION WITH EMBEDDED ELECTROFUSION WIRE

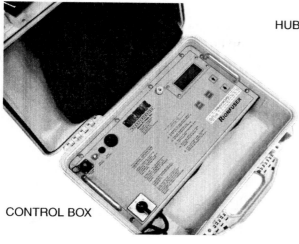

CONTROL BOX

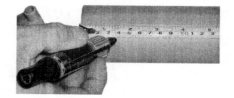

Step 1. Mark the coupling depth on the pipe or fitting, using manufacturer-supplied information for specific pipe sizes.

Step 2. Sand the shiny surface from the area of the pipe to be fused.

Step 3. Insert the pipe completely into the coupling. Check that it meets the measured mark around the pipe.

Step 4. Repeat these steps with the fitting. Both ends of the coupling will fuse at the same time.

Step 5. Connect the leads to the coil terminals.

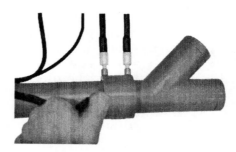

Step 6. Enter the appropriate settings on the control box and press the "start" button on either the output lead or the control box.

Figure 12 ◆ The fusion method. (2 of 2)

The other methods of joining plastic pipe are familiar to you, including using solvent cement to join thermoplastic pipe. Always follow the manufacturer's instructions carefully. Most codes permit mechanical joints (see *Figure 13*) on pipes that are installed underground. Some types of plastic pipe can also be threaded and joined using fittings such as elbows, tees, wyes, adapters, and reducers. Your local code will specify the preferred methods of joining thermoplastic pipe.

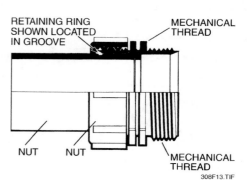

Figure 13 ◆ Mechanical joint for thermoplastic pipe.

Thermoset pipe can be joined with threading or mechanical joints, but the most common way to join thermoset pipe is to use an epoxy **adhesive**. Adhesives are chemicals that bond surfaces without fusing them together. Use them on pipes that have tapered bell and spigot fittings (see *Figure 14*). Apply adhesives the same way you would apply solvent cement. Note that most pipes and adhesives are not interchangeable among manufacturers.

When you install plastic pipe, follow your local code. Building plans will describe the proposed system. Avoid hangers that may cut or squeeze the pipe and tight clamps that prevent the pipe from moving or expanding. Support the pipe at intervals of no more than 4 feet and at branches, at changes of direction, and when you use large fittings. Do not install plastic pipe near sources of excessive heat or vibration. Avoid installing the hanger near sharp objects that could rub and cut the pipe.

WARNING!
Use protective eyewear and gloves when you are using solvent cement. Avoid skin or eye contact with the cement. If contact occurs, wash the affected area immediately. Always apply cement in a well-ventilated area.

4.3.0 Joining and Installing Silicon Cast-Iron Pipe

As mentioned earlier, silicon cast-iron pipe comes in hub-and-spigot and no-hub styles. Consult the waste system's design drawings to find out which type is preferred for the installation. The procedure for joining hub-and-spigot silicon cast-iron pipe is similar to that used for ordinary iron waste pipe. The materials are specially designed to resist corrosion. Use the following steps to join hub-and-spigot pipe (refer to *Figure 15*):

Step 1 Insert the spigot end of one pipe into the hub end of the other pipe. Leave a slight gap between the end of the spigot and the bottom of the hub. The gap prevents corrosion caused by contact between the two metals. It also prevents pipe strain during expansion.

Step 2 Carefully ram acid-resistant rope into the bottom of the hub. Fill the hub approximately half full.

Step 3 Pour lead into the hub until it is filled.

WARNING!
When you are pouring lead, wear safety glasses, gloves, high-top shoes or boots, and long pants. These items will protect you from splattering hot lead.

Step 4 Caulk the lead the same way you would caulk cast-iron soil pipe. Take care not to break the pipe while you are caulking.

WARNING!
Moisture can cause molten lead to explode out of a joint, resulting in serious injuries. Ensure that the pipe is dry before you pour lead.

ON THE LEVEL — Cold Weather Tip

In colder climates, preheat joints before pouring lead into them, to avoid splitting the joint due to thermal expansion.

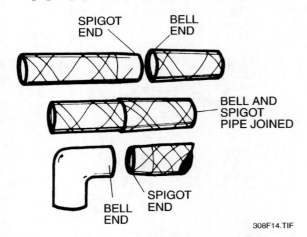

Figure 14 ◆ Examples of thermoset plastic pipe with bell and spigot fittings.

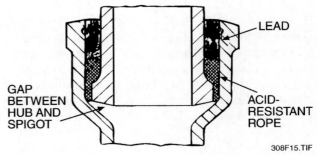

Figure 15 ◆ Corrosive-resistant joint in hub-and-spigot silicon cast-iron piping.

No-hub silicon cast-iron pipe is relatively easy to assemble. Use mechanical joints to assemble pipe and fittings. Review the manufacturer's instructions to ensure that the joints are suitable for the type of pipe being installed. Use the following procedure to join pipes and fittings with mechanical joints (see *Figure 16*):

Step 1 Slide the joint sleeve over the end of the pipe until it reaches the positioning lug on the pipe.

Step 2 Slide the other pipe or fitting into the joint.

Step 3 Tighten the two bolts on the joint. The torque on the bolts should be at least 9 foot-pounds.

Hub-and-spigot and no-hub silicon cast-iron pipe can be installed underground or in concrete without special preparation. For underground installations, ensure that the first few inches of fill are free of rocks. Some mechanical joints require an asphalt mastic coating if they are installed on underground pipe. Refer to the manufacturer's instructions.

Install silicon cast-iron pipe using the same methods you would use for regular cast-iron soil pipe. Support each length of pipe in a horizontal run. If the line includes fittings, place the supports no farther than 7 feet apart. Do not completely tighten supports around the horizontal pipe because the pipe should be free to expand and contract. Install supports directly beneath fittings that connect to vertical stacks.

Support vertical runs at least every 14 feet. Refer to the manufacturer's instructions for approved hangers. Install suitable cleanout plugs on iron waste piping systems. Clean waste systems using a sewer auger or chemical cleaner that is safe for use in silicon cast-iron pipe.

4.4.0 Joining and Installing Stainless Steel Pipe

Stainless steel corrosive-resistant waste pipe can be welded. It is usually too thin to be joined by threading. Do not attempt to weld steel pipe unless you have been trained in proper welding techniques. Always use appropriate personal protective equipment when you are working with welding tools. Ensure that the weld rod has the same alloy content as the pipe metal. Otherwise, the weld joints could be weakened through contact with the corrosive waste. Note that different welding techniques apply to the individual steel alloys.

Install stainless steel pipe as you would other types of metal piping. Ensure that hangers and

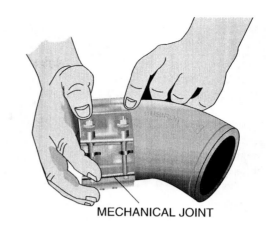

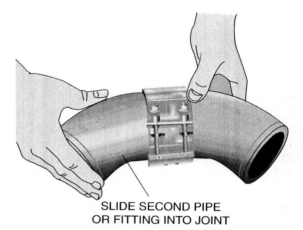

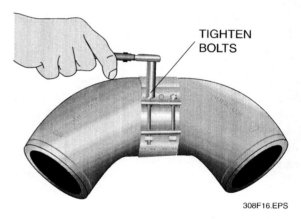

Figure 16 ♦ Installing a mechanical joint on a no-hub pipe.

supports are properly sized and located according to your local code. Provide additional support for fittings and other accessories on the line. When you are installing insulated pipe, you are required to place hangers and supports closer together. Never spring stainless steel pipe into place. Springing causes stress that will seriously weaken the pipe. External loads on an unsupported length of pipe will also cause stress fractures. Always follow the manufacturer's directions when installing stainless steel pipe.

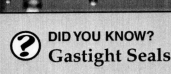

DID YOU KNOW?
Gastight Seals

When you are dealing with contaminated stainless steel, you can use a lightweight crimping tool to create nearly gastight seals. These seals can be used in contaminated stainless steel tubing up to 1 inch in diameter. The crimper removes the contaminated section and provides a uniform seal across the crimped surface.

4.5.0 Installing Acid Dilution and Neutralization Sumps

Corrosive liquids, spent acids, and toxic chemicals pose serious health risks. They cannot be discharged directly into the drainage system. They must first be diluted, chemically treated, or neutralized. Install acid dilution and neutralization sumps to treat corrosive wastes. Consult your local code for specific requirements. Many codes require that you obtain approval before you install acid dilution and neutralization sumps.

If the waste flow is intermittent, it can be diluted. Ensure that the sump is large enough to permit the waste to dilute to an acceptable level according to local code. Refer to your local code for chemicals that can be treated through dilution. Sumps used for dilution must be vented (see *Figure 17*). Chemical waste and sanitary vents must extend through the roof. Some dilution tanks may be vented separately; refer to the manufacturer's instructions and to your local applicable code. Never vent a dilution tank into the sanitary system.

Some wastes must pass through a neutralizing agent before they can be discharged (see *Figure 18*). Neutralizing agents react chemically with corrosive wastes. The result is a waste product that can be discharged safely. One of the most commonly used agents is appropriately sized limestone or other neutralizing material. Stones that are too small will restrict the flow. Install vents and gas-tight lids on sumps with neutralizing agents. Ensure that the sump contains enough of the neutralizing agent to treat all of the special waste that will flow through it.

Install the sump where it is easily accessible. Change the neutralizing agent regularly. The manufacturer?s instructions will specify how often it should be changed. Allow only trained professionals to service a sump. Take proper precautions when you open a sealed sump lid, and seal the lid tightly when you are finished. Refer to your local code for specific guidelines on servicing sumps.

5.0.0 ◆ HAZARD COMMUNICATION

A recent government survey found that 650,000 hazardous chemicals are commonly used in workplaces all around the country. Each of them can pose health and safety risks if not handled properly. Workers have a right to know what the risks are. Employers develop **hazard communication** (HazCom) plans to educate and protect their employees. HazCom programs are required by law. As part of a HazCom program, an employer must do the following:

- Post lists of hazardous chemicals in the workplace.
- Create a written safety plan.
- Put labels or tags on all chemical containers.
- Keep material safety data sheets (MSDSs) on each of the hazardous chemicals that are used or stored.
- Train workers in proper safety and handling techniques.

HazCom programs vary depending on the types of chemicals that are used and stored. Be sure to find out about your company's HazCom program. Your safety and that of your co-workers depend on it.

5.1.0 HazCom Labels

The National Fire Protection Association (NFPA) has developed labeling guidelines for hazardous chemicals. The **NFPA diamond** is one of the most commonly used HazCom labels (see *Figure 19*).

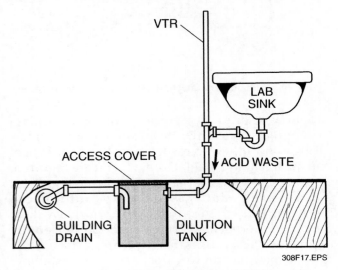

Figure 17 ◆ Venting a dilution tank through the roof.

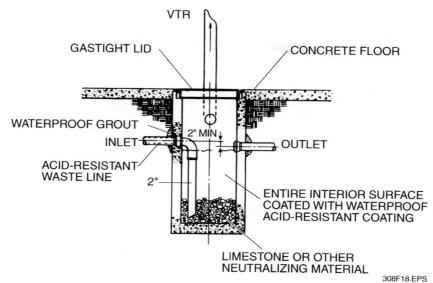

Figure 18 ◆ Neutralizing sump with neutralizing agent.

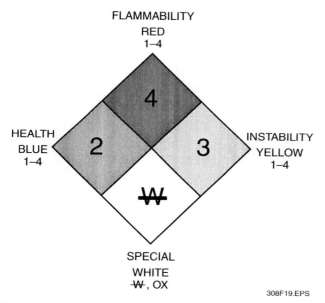

Figure 19 ◆ The NFPA diamond.

You may have seen this label hanging on walls where you work. The diamond serves as a quick reference to the nature of the chemicals inside. The diamond is divided into four quarters. Numbers in each quarter indicate one of the properties of the chemical (see *Table 1*). With a little practice, you will learn to decipher a diamond label quickly.

Post the labels near all doors leading into an area where chemicals are stored or used. Ensure that labels are posted by outside entrances where rescue workers can see them. The NFPA's labeling standard is designed to protect workers and the public. It is also intended to help response teams react swiftly in an emergency. The labels do not include detailed information about the types of chemicals being stored. *NFPA Standard 704* describes how, where, and when to use the labeling system. You can order a copy of the standard from the NFPA website, www.nfpa.org.

Laboratories use special signs to indicate potential hazards. The **Uniform Laboratory Hazard Signage** (ULHS) system is widely used in many different types of laboratories. ULHS signs are standards that are used around the world. They indicate the type of risk by using special graphics called pictographs (see *Figure 20*). Signs warn against oxidizers, electrical hazards, cancer hazards, **biohazards**, and other risks. A biohazard is an organism or chemical that poses a health risk to humans, animals, or plants. Refer to your local code before posting laboratory signs.

5.2.0 Material Safety Data Sheets

MSDSs are bulletins that contain a wide range of useful safety information. Manufacturers are required to create MSDSs for each of their chemical products. Companies must keep MSDSs on file for all of the chemicals that are used or stored on site. The MSDSs should be kept where they can be found easily. All employees must know where to find MSDSs.

An MSDS lists the chemical and common names of a product as well as the health and safety hazards associated with it. If the chemical is a compound, the names of dangerous ingredients are listed. The MSDS also describes the product's boiling and freezing points, density and specific gravity, appearance, and odor. If the product could ignite or explode, the MSDS will list the conditions that would present a danger. Chemicals that react with the product are also listed.

Table 1 Components of the NFPA Diamond Label

Color	Location	Meaning	Severity
Red	Top	Flammability Hazard	0 = Will not ignite 1 = Will ignite if preheated 2 = Will ignite if moderately heated 3 = Will ignite at normal temperatures 4 = Extremely flammable
Blue	Middle Left	Health Hazard	0 = No health hazard 1 = Slight health hazard 2 = Requires use of breathing apparatus 3 = Requires complete appropriate personal protective equipment 4 = Extreme health hazard—avoid any contact
Yellow	Middle Right	Instability Hazard	0 = Stable 1 = Unstable if heated 2 = Violent chemical reaction possible 3 = Will detonate by strong shock or heat 4 = May detonate—evacuate if fire risk is present
White	Bottom	Special Hazard	W̶ = Reacts if exposed to water OX = Reacts if exposed to oxygen

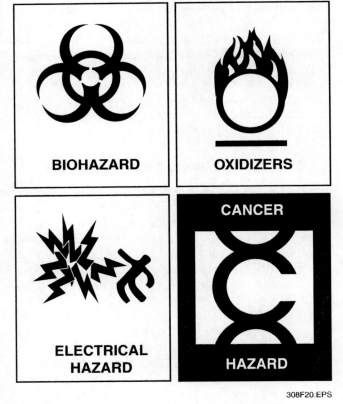

Figure 20 ♦ Examples of laboratory hazard signs.

The MSDS provides instructions for safe handling and use of the product. It also describes what to do in an emergency. Always consult the MSDS before you work with a chemical product. Never dispose of a chemical product without consulting the MSDS or your safety officer. Refer to your workplace health code for more information.

5.3.0 Worker Training

Training programs are an effective way to ensure that employees understand the company's HazCom program. Train new employees when they join the company. Conduct annual reviews for all employees who work with hazardous chemicals. The training should cover the following topics:

- Where to find information such as MSDSs and the written safety plan
- How to detect the release of chemicals
- How to identify health and physical hazards
- How to protect yourself and others in the event of a chemical release

All employees must know the safety regulations and where dangerous chemicals are stored and used. Study your company's HazCom plan. Don't be afraid to ask questions about the details of the plan. In an emergency, this knowledge could save your life and the lives of your co-workers.

Review Questions

Sections 4.0.0–5.0.0

1. When you join borosilicate glass piping, the first step is to _____.
   a. snap the coupling over the pipe bead
   b. seat the plain end
   c. wet the inside of the coupling
   d. tighten the coupling bolt

2. Use each of the following methods/materials to join thermoplastic pipe except _____.
   a. pipe threading
   b. epoxy adhesives
   c. mechanical joints
   d. fusion

3. In a horizontal run of silicon cast-iron pipe with fittings, place the supports no farther than _____ feet apart.
   a. 3
   b. 5
   c. 7
   d. 9

4. In the NFPA diamond label, the middle right square indicates a(n) _____ hazard.
   a. instability
   b. health
   c. flammability
   d. special

5. MSDS is an abbreviation for _____.
   a. master safety data sheet
   b. material safety data standard
   c. master safety data system
   d. material safety data sheet

Summary

Plumbers install DWV systems that are designed to handle corrosive wastes. Corrosive wastes are extremely acidic or alkaline chemicals. They can damage pipes and cause illness, injury, or death. Corrosive wastes must be disposed of through a separate drainage system after being treated. This module discussed how to install corrosive-resistant waste piping in commercial facilities.

The most common materials used in corrosive waste pipe and fittings are borosilicate glass, thermoplastic and thermoset plastics, silicon cast iron, and stainless steel. Each of these materials is used to handle different types of corrosive chemicals.

Borosilicate glass pipe is made from a specially reinforced transparent glass. It is used in systems with high-temperature wastes. Thermoplastic and thermoset pipe can handle acetic acids and chemical compounds containing aluminum, ammonia, barium, calcium, sodium, and mild sulfuric acids. Resistance will vary depending on the type of pipe. Silicon cast-iron pipe can handle almost any type of corrosive waste. Stainless steel pipe is available in four types of alloys for corrosive-resistant waste piping. The four alloys are ferritic, austenitic, martensitic, and duplex. Each type offers its own special corrosion resistance capabilities.

Install corrosive-resistant waste systems using techniques that you have already learned. Specific requirements depend on the type of pipe. Glass pipe can be installed both horizontally and vertically. Use padded hangers for horizontal runs of glass pipe. Fit a protective casing on glass piping that is installed underground. Thermoplastic and thermoset pipe can be joined by fusion, solvent, mechanical joints, or threading. Do not use hangers that may cut or squeeze the pipe. Silicon cast-iron pipe comes in hub-and-spigot and no-hub styles. Install acid dilution sumps according to local code. Stainless steel pipe may be welded. Different welding techniques apply to each steel alloy.

Safe drainage is not the only way to protect people from chemicals. This module also addressed how hazard communication plans help prevent chemical accidents. Employers must post a list of all chemicals that they use or store and create a written safety plan in case of an emergency. They must label or tag all chemical containers and keep MSDSs for all chemicals on site. Employers must also offer regular safety training. These steps will help to ensure a safe working environment.

Notes

Trade Terms Introduced in This Module

Adhesive: A chemical used to bond thermoset pipes and fittings without fusing them together.

Austenitic stainless steel: A stainless steel alloy that contains nickel and stays strong at very low temperatures. Austenitic steel is the most popular type of steel for corrosive waste pipes.

Beaded end: A raised lip around the rim of borosilicate glass pipe.

Biohazard: An organism or chemical that is dangerous to the health of living creatures.

Borosilicate glass: A pipe material made from specially reinforced glass that does not expand or contract much with changes in temperature.

Corrosive waste: Waste products that contain harsh chemicals that can damage pipe and cause illness, injury, or even death.

Coupling: Removable metal clamps with plastic seals used to connect borosilicate glass pipe.

Duplex stainless steel: A stainless steel alloy mix of austenitic and ferritic stainless steels that offers very good corrosion resistance.

Ferritic stainless steel: A stainless steel alloy that is very resistant to rust and has a very low carbon content.

Furan: A resin that is highly resistant to solvents. Furans are used to make thermoset pipe.

Fusion coil: A device that uses heat to join thermoplastic pipe.

Hazard communication: A workplace safety program that informs and educates employees about the risks from chemicals.

Martensitic stainless steel: A stainless steel alloy that is very strong but is mostly used to handle mild corrosive wastes.

NFPA diamond: A label that uses color-coding and numbers to indicate the properties of chemicals. The National Fire Protection Association developed the diamond symbol.

Thermoplastic: A type of plastic pipe that can be softened and reshaped when it is exposed to heat.

Thermoset: A type of plastic pipe that is hardened into shape when manufactured.

Uniform Laboratory Hazard Signage: Signs placed on laboratory walls that use pictographs to indicate hazards.

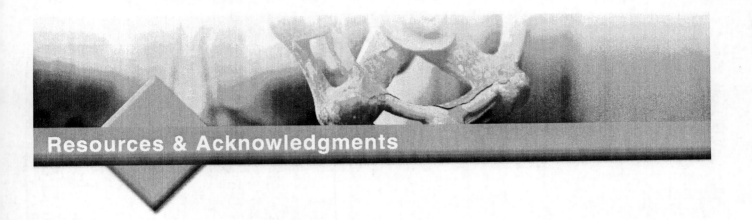

Resources & Acknowledgments

Additional Resources

This module is intended to be a thorough resource for task training. The following reference works are suggested for further study. These are optional materials for continued education rather than for task training.

Corrosion-Resistant Piping Systems. 1994. Philip A. Schweitzer. New York: Marcel Dekker, Inc.

Hazard Communication Made Easy: A Checklist Approach to OSHA Compliance. 2000. Sean M. Nelson and John R. Grubbs. Rockville, MD: ABS Group, Government Institutes.

Water, Sanitary, and Waste Services for Buildings. 2002. Alan F. E. Wise and J. A. Swaffield. Boston, MA: Addison-Wesley.

Figure Credits

Flowserve Corporation, 308F06

Orion Fittings, Inc., 308F12

Schott North America, Inc., 308F02, 308F04

References

2003 International Plumbing Code. 2003. International Code Council. Falls Church, VA: International Code Council.

Cast Iron Soil Pipe Institute Website, http://www.cispi.org, Cast Iron Soil Pipe & Fittings Handbook, http://www.cispi.org/handbook.htm, reviewed July 2005.

Corrosion-Resistant Piping Systems. 1994. Philip A. Schweitzer. New York: Marcel Dekker, Inc.

Dictionary of Architecture and Construction, Third Edition. 2000. Cyril M. Harris, ed. New York: McGraw-Hill.

Flowserve Corporation Website, http://www.flowserve.com, *Duriron Bell and Spigot Pipe and Fittings*, Bulletin PF-9, http://www.flowserve.com/other/pdfs/bulletin_PF9.pdf, reviewed July 2005.

Oklahoma State University, Department of Environmental Health and Safety Website, http://www.pp.okstate.edu/ehs, *Hazard Communication Training Manual*, http://www.pp.okstate.edu/ehs/HAZCOM/Manual.htm, reviewed July 2005.

Pipefitters Handbook. 1967. Forrest R. Lindsey. New York: Industrial Press Inc.

Planning Drain, Waste & Vent Systems. 1993. Howard C. Massey. Carlsbad, CA: Craftsman Book Company.

United States Army Corps of Engineers, Publications Web site, http://www.usace.army.mil/publications, Engineer Manual EM 1110-1-4008, *Engineering and Design: Liquid Process Piping*, http://www.usace.army.mil/publications/eng-manuals/em1110-1-4008/toc.htm, reviewed July 2005.

Water, Sanitary, and Waste Services for Buildings. 1995. Alan F. E. Wise and J. A. Swaffield. Boston, MA: Addison-Wesley.

CONTREN® LEARNING SERIES — USER UPDATE

The NCCER makes every effort to keep these textbooks up-to-date and free of technical errors. We appreciate your help in this process. If you have an idea for improving this textbook, or if you find an error, a typographical mistake, or an inaccuracy in NCCER's Contren® textbooks, please write us, using this form or a photocopy. Be sure to include the exact module number, page number, a detailed description, and the correction, if applicable. Your input will be brought to the attention of the Technical Review Committee. Thank you for your assistance.

Instructors – If you found that additional materials were necessary in order to teach this module effectively, please let us know so that we may include them in the Equipment/Materials list in the Annotated Instructor's Guide.

Write: Product Development and Revision
National Center for Construction Education and Research
P.O. Box 141104, Gainesville, FL 32614-1104

Fax: 352-334-0932

E-mail: curriculum@nccer.org

Craft _____ Module Name _____

Copyright Date _____ Module Number _____ Page Number(s) _____

Description
_____
_____
_____
_____

(Optional) Correction
_____
_____
_____

(Optional) Your Name and Address
_____
_____
_____

Plumbing Level Three

02309-06

Compressed Air

02309-06
Compressed Air

Topics to be presented in this module include:

1.0.0	Introduction	9.2
2.0.0	Working Safely with Compressed Air	9.2
3.0.0	Principles of Compressed Air Systems	9.3
4.0.0	Components of Compressed Air Systems	9.5
5.0.0	Installing Compressed Air Systems	9.10

Overview

Compressed air is used to operate a wide variety of tools, machines, and equipment. Compressed air systems are used in numerous industries. Many systems are portable. Many others are permanently installed in buildings such as factories, to supply a large number of machines or tools with a constant supply of compressed air. Plumbers are responsible for installing these systems.

The major components of a compressed air system are the air compressor and related equipment, aftercoolers and (sometimes) air dryers, supply piping, and controls for the compressor and the distribution system. Air compressors condense air to the desired pressure and distribute it throughout the system. The most common types are reciprocating compressors and rotary compressors. Since both excess heat and moisture can damage the system, aftercoolers cool and dry the air by passing it through a heat exchanger before distribution. For some applications that require air without a trace of moisture, plumbers install air dryers. A variety of controls, including switches, throttles, and valves, keep the system pressure within the control range (the range between the highest and lowest allowable air pressure).

A good compressed air system provides the right amount of high-quality air at the correct pressure. To troubleshoot low pressure, a common problem in compressed air systems, plumbers test for leaks and evaluate pipe sizing and compressor capacity. Plumbers must always take safety precautions when working with compressed air, by wearing personal protective equipment, disconnecting power sources before maintenance, and securing hoses and connections.

Focus Statement
The goal of the plumber is to protect the health, safety, and comfort of the nation job by job.

Code Note
Codes vary among jurisdictions. Because of the variations in code, consult the applicable code whenever regulations are in question. Referring to an incorrect set of codes can cause as much trouble as failing to reference codes altogether. Obtain, review, and familiarize yourself with your local adopted code.

Portions of this publication reproduce tables and figures from the *2003 International Plumbing Code* and the *2000 IPC Commentary*, International Code Council, Inc., Falls Church, Virginia. Reproduced with permission. All rights reserved.

Objectives

When you have completed this module, you will be able to do the following:

1. Identify the components of compressed air systems.
2. Discuss the installation of compressed air systems and their components and accessories.
3. Describe the applications of compressed air systems.
4. Identify the different methods of conditioning compressed air.
5. Identify the types, functions, and capacities of different air compressor systems.
6. Identify the safety issues related to compressed-air systems.
7. Install a basic compressed-air system.

Trade Terms

Actual cubic feet per minute
Aftercooler
Air dryer
Air end
Air pressure regulator
Air receiver
Air throttle valve
Boyle's law
Compressible
Constant speed control
Control range
Coupling
Desiccant air dryer
Hydraulics
Intercooler
Lubricator
Network control
Part-load control
Pneumatics
Reciprocating compressor
Refrigeration air dryer
Rotary compressor
Rotary screw
Sequencing control
Standard conditions
Standard cubic feet per minute
Stop/start control
Throttle
Unloading
Use factor
Water drip leg

Required Trainee Materials

1. Appropriate personal protective equipment
2. Sharpened pencils and paper
3. Copy of local applicable code
4. Calculator

Prerequisites

Before you begin this module, it is recommended that you successfully complete *Core Curriculum; Plumbing Level One; Plumbing Level Two; Plumbing Level Three*, Modules 02301-06 through 02308-06.

This course map shows all of the modules in the third level of the *Plumbing* curriculum. The suggested training order begins at the bottom and proceeds up. Skill levels increase as you advance on the course map. The local Training Program Sponsor may adjust the training order.

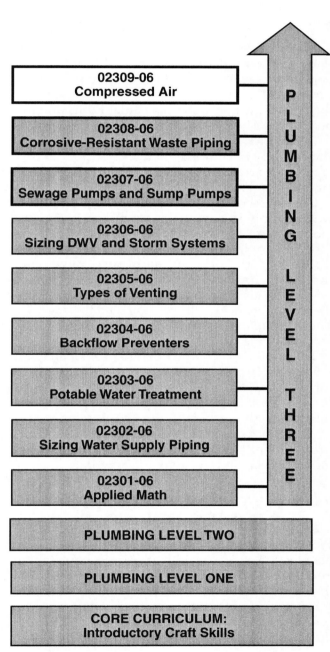

1.0.0 ♦ INTRODUCTION

Compressed air is used to operate a wide variety of tools, machines, and equipment. Factories, electronics manufacturers, laboratories, automobile detailers, and construction companies are just some of the places where compressed air systems are used. The following are some types of equipment that use compressed air:

- Construction tools and equipment
- Paint sprayers and powder-coating machines
- Respirators to protect against inhaled fumes and particles
- Air supply and climate control systems

Many compressed air systems are portable. They are completely self-contained and can be hooked up on site. When they are not in use, they can be stored for long periods of time. Other compressed air systems are permanently installed in buildings. They are designed to serve a larger number of tools and machines or to provide a constant supply of compressed air. This module covers the design and installation of large compressed air systems for buildings.

Compressed air systems work very much like water supply systems. Air is supplied to the system by an air compressor, which serves the same function as a water pump. The air is circulated through a series of mains and risers. Valves control the airflow. The fixtures in a compressed air system are air-powered tools and machines. Keep these similarities in mind as you read this module. They will help you grasp the concepts quickly and accurately.

This module is not intended to provide instruction on the installation of compressed air systems for hospitals or medical facilities. The systems used in those facilities are specialized and require training in the installation of those types of systems. Refer to your local applicable code and the professional training and certification requirements that are in force where you work.

2.0.0 ♦ WORKING SAFELY WITH COMPRESSED AIR

Working with compressed air can present a number of safety risks. As with all aspects of installing and servicing plumbing systems, you should maintain good safety practices when installing and working with compressed air systems.

Compressed air can damage tools and equipment or cause injury if not properly handled. Before operating tools that use compressed air, always refer to the manufacturer's manual and become familiar with any specific operating procedures and safety rules. And remember, when using compressed air and air powered tools, to consider the safety of any co-workers that are in the proximity of your work.

Compressed air can be dangerous because of its high temperature and high air pressure. Air compressors can become very hot and can cause serious burns, and accidental release of compressed air can cause injury or death. Remember to exercise caution in the following ways:

- Wear appropriate personal protective equipment when using any air-powered tool.
- Never place a compressed air source in or near your nose, mouth, eyes, or skin.
- Wear proper hearing protection. The very high pitch of some air-operated tools could cause hearing damage or loss.
- Keep your hands away from the rotary components of a compressed air system.
- Ensure that all connections are secure before activating an air-powered tool or machine. An insecure connection can cause a hose to whip around with tremendous force.

Take suitable safety precautions when installing, repairing, or servicing air compression systems. Always disconnect the power source for any tool before you replace parts such as bits, blades, or discs. Always disconnect the power source before you perform maintenance on any power tool, and never bypass the trigger lock on any power tool.

Use caution during installation of piping for compressed air systems. Wear appropriate personal protective equipment when installing threaded pipe. The chips made from threading pipe are sharp and could cut you. Always use a rag to wipe off excess oil; never use your bare hands. In addition, always ensure that the inside of the pipe is clean before you install it. Debris inside the pipe will become projectiles when accelerated by the air, and can damage equipment and cause serious injury.

Finally, protect compressed air tools and equipment from damage. When using any air-operated tool or machine, make sure that all hoses and couplings are secured. Otherwise, the pressure may be too low for the tool or machine to operate properly. The result could damage the device. Also, before you service a compressed air system, shut off the system and allow the pressure to escape from all components and air lines.

3.0.0 ♦ PRINCIPLES OF COMPRESSED AIR SYSTEMS

Much of your training has focused on systems that circulate or dispose of water and liquid wastes. You understand concepts such as temperature, pressure, density, and friction. These concepts all have to do with **hydraulics**, the set of rules and principles that governs the motion of water and other liquids.

Systems that use compressed air instead of water rely on a different set of rules called **pneumatics**. The rules and principles of pneumatics govern the motion of air and other gases. Temperature, pressure, density, and friction play important roles in pneumatics.

Air pressure is an important principle in pneumatics. You know that air pressure plays a key role in many plumbing operations by:

- Equalizing pressure in drainage pipes through vents
- Absorbing shock waves inside water hammer arresters
- Providing power for air hammers, nail guns, and other tools

Air pressure is discussed in more detail below. There are many similarities and some important differences between water and air. Learning the basics of pneumatics will help you understand how compressed air systems work.

3.1.0 Properties of Air

Unlike water, air is **compressible**. It can be squeezed to fit into a smaller space than it occupies at normal air pressure. This increases its pressure. The higher the air's pressure, the more force it applies to the walls of the surrounding container. Storage tanks and piping have to be strong enough to withstand the air pressure in the system. Always check the pressure rating of materials and equipment before you install them.

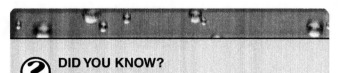

DID YOU KNOW?

The word hydraulic comes from the ancient Greek word hydraulis, which was a water-powered musical instrument. The word pneumatic comes from another ancient Greek word pnein, which means to sneeze.

DID YOU KNOW?

More than 1,500 years ago, a Greek mathematician and inventor named Hero described how water and compressed air could be used to power machines. He wrote a book called *The Pneumatics* that explained his ideas in detail. One of his inventions used compressed air to force a jet of water through a nozzle. Another device channeled steam from boiling water into a small sphere held by two pivots. As the steam escaped through two bent tubes on the sphere, it caused the sphere to rotate. This device is one of the first recorded descriptions of a steam engine.

When air is compressed, its temperature increases, and it gives off heat. When it expands, it absorbs heat from its surroundings. Consequently, compressed air systems are designed to protect operators from temperature extremes. Compressors have insulation to prevent operators from touching a hot surface. Pneumatic tools are designed to protect operators from the cold. Learn how to handle compressed air tools properly. Wear appropriate personal protective equipment when working with compressed air systems and tools.

3.2.0 Measuring Air Pressure

Compressed air pressure is measured in pounds per square inch gauge (psig). Psig uses local air pressure as the zero point. At sea level, 0 psig is 14.7 pounds per square inch absolute (psia). The flow rate of compressed air is measured in cubic feet per minute (cfm).

The air inlet capacity of an air compressor can be expressed in terms of two forms of cfm: either **standard cubic feet per minute** (scfm) or **actual cubic feet per minute** (acfm). Scfm is the volume of air measured at standard conditions. The following are considered **standard conditions**:

- A sea-level air pressure of 14.7 psia
- A temperature of 68°F
- A barometric pressure of 29.92 inches of mercury

On the other hand, acfm is a measure of the air volume under the actual temperature and pressure conditions at the inlet of the air compressor.

The acfm of an air compressor remains constant regardless of the ambient temperature and pressure.

To convert measurements from scfm to acfm and vice versa, use **Boyle's law**. Boyle's law states that if the air pressure in a system decreases, the air volume will increase proportionally:

$$P_1V_1 = P_2V_2$$

Where:

P = Pressure
V = Volume

The calculation is performed using 29.92 inches of mercury as an absolute value for P_1.

Consider the following example: You need to find the equivalent of 35 scfm for an air compressor with an inlet vacuum level of 27 inches of mercury. The air compressor's inlet temperature is 68°F, and the compressor is located at sea level. Begin by converting the given inlet vacuum value to an absolute figure.

$$P_2 = 29.92 - 27$$
$$P_2 = 2.92$$

Now substitute the given numbers for the variables in Boyle's law and perform the math:

$$P_1V_1 = P_2V_2$$
$$29.92 \times 35 \text{ scfm} = 2.92 \times V_2 \text{ acfm}$$
$$1047.2 \text{ scfm} = 2.92 \times V_2 \text{ acfm}$$
$$V_2 \text{ acfm} = \frac{1047.2}{2.92}$$
$$V_2 \text{ acfm} = 358.63$$

The equivalent air volume for the air compressor is 358.63 acfm.

Review Questions

Sections 2.0.0–3.0.0

1. Because of its high _____, accidental release of compressed air can cause injury or death.
   a. temperature
   b. pressure
   c. acfm
   d. moisture content

2. If all hoses and couplings of an air-operated tool are not secured, the tool may be damaged because the _____.
   a. system has no ventilation
   b. volume is less than the pressure
   c. air is too hot
   d. pressure is too low

3. The higher the _____ of compressed air, the more force it applies to the walls of the surrounding container.
   a. humidity
   b. pressure
   c. temperature
   d. velocity

4. Air compressor inlet capacity can be expressed in terms of _____.
   a. psia
   b. psig
   c. acfm
   d. gpm

5. Boyle's law states that if the air pressure in a system decreases, the _____ will increase proportionally.
   a. air volume
   b. air temperature
   c. psig
   d. air vacuum

4.0.0 ♦ COMPONENTS OF COMPRESSED AIR SYSTEMS

A compressed air system collects, prepares, and distributes air to a variety of tools and machines (see *Figure 1*). The following are the major components of a compressed air system:

- Air compressors and related equipment
- **Aftercoolers** and **air dryers**
- Supply piping
- Controls

Each of these components is discussed in greater detail in the following sections. Review how each one works, both on its own and in combination with the other components. Your local code will have specific requirements for components that can be used in your area.

4.1.0 Air Compressors

Air compressors condense air to the desired pressure and distribute it throughout the system. Air compressors come in a variety of sizes. If you have worked on a construction site, you have probably seen a portable air compressor like the one in *Figure*

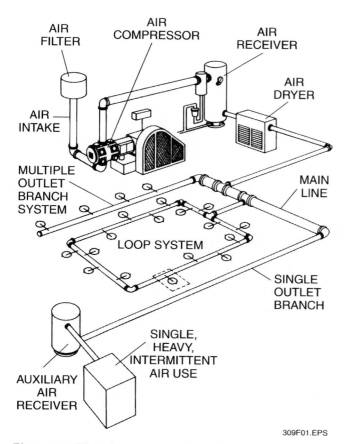

Figure 1 ♦ Typical compressed air system.

Figure 2 ♦ Portable air compressor.

2. Many trades use portable air compressors to operate pneumatic tools. Plumbers use portable air compressors for pressure tests on water supply systems. On the other end of the scale are compressors that supply air to an entire factory.

Air pressures also vary widely. Some air compressors used in precision work produce 10 psig or less. Air compressors used in wind tunnels, by contrast, produce more than 2,500 psig. In the average building system, air compressors generally operate at pressures of 500 psig or lower. Most air compressors operate between 90 and 125 psig.

Reciprocating and **rotary** are the most common types of air compressors. In reciprocating compressors, air is compressed by a piston (see *Figure 3*). The compression stroke pumps the air into the supply line. Rotary compressors use **rotary screws** to compress the air (see *Figure 4*). Rotary screws work by compressing air as it travels down the screw path. The air is directed into a chamber, called the **air end**, at the end of the screws. As more air collects there, it is compressed further.

Electric motors or gasoline engines provide the power for most air compressors. Centrifugal air compressors are also available. All compressors have air filters on the inlet side.

From the compressor, the air flows into an **air receiver**, which acts as a storage tank and also smoothes out the airflow from a reciprocating compressor. Water that was trapped in the air while it was being compressed condenses out in the air receiver. Many air receivers have automatic drains to remove the water from the tank. Other tanks must be drained manually. Relief valves allow air receivers to vent excess pressure (see *Figure 5*). Codes require that all air receivers and relief valves meet the safety standards of the American Society of Mechanical Engineers (ASME).

Figure 3 ◆ Reciprocating compressor with piston.

Multistage air compressors have more than one compressor. Each stage is a complete compressor. Each stage passes air on to the next one to be compressed further until the desired pressure is reached. Multistage air compressors can deliver a larger volume of air than single-stage units. Multistage compressors are usually equipped with an **intercooler**. Intercoolers are heat exchangers that remove heat from compressed air as it moves from the lower pressures of the early stages to the higher pressures of the later stages. The number of stages depends on the volume demands of the compressed air system.

Air compressors, motors, and air receivers can be arranged to fit the available space. The motor and air compressor can be mounted on top of the storage tank (see *Figure 6*). Larger systems are often installed with the air storage tank separate from the air compressor and motor.

Air compressors tend to make noise and vibrate while operating. You can install a variety of devices to reduce noise and vibration. Intake silencers quiet the flow of air into the compressor.

Figure 5 ◆ Air receiver relief valve.

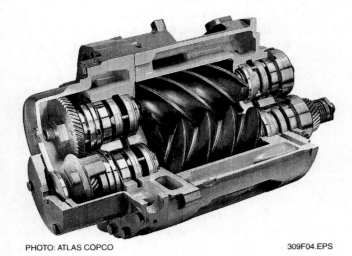

Figure 4 ◆ Rotary screw element in a rotary compressor.

Figure 6 ◆ Air compressor with motor and compressor mounted on top of the air receiver.

Flexible metal hoses prevent the transmission of vibrations from the compressor to the rest of the system (see *Figure 7*). Shock-absorbing mounts isolate the unit from the floor.

4.2.0 Aftercoolers and Air Dryers

It takes 7 cubic feet of air to make 1 cubic foot of compressed air at 100 psig. However, the moisture in the air is compressed along with the air. That means that the compressed air has seven times the moisture of free air. Water vapor in the air can cause rust inside the compressed air system and inside the machines that use the air. In addition, compressed air is hotter than free air. Excess heat could damage the system. Before compressed air can be used, therefore, it has to be dried and cooled.

Compressed air systems use aftercoolers to cool and dry the air. An aftercooler cools the air by passing it through a heat exchanger. The drop in temperature causes the moisture in the air to condense. The condensed moisture precipitates out of the air and drains away. Aftercoolers may use either air or water to cool the air (see *Figure 8*).

Aftercoolers remove most of the moisture from compressed air, but a small amount can remain.

ON THE LEVEL
Small Electric Motors and Air Compressors

Air compressors that are powered by small electric motors (under 25 horsepower) usually stop running between cycles. Install an air receiver in the system to allow the compressor to stop and start as required. When the air pressure in the air receiver reaches a preset maximum, a pressure switch turns off the motor. When the air pressure drops below a preset minimum, the switch turns on the motor again.

Certain applications require air with no trace of moisture. In such cases, air dryers are installed to remove all moisture. **Refrigeration air dryers** and **desiccant air dryers** are the most common types.

Refrigeration units pass air through a heat exchanger. A liquid refrigerant is also pumped through the heat exchanger. The refrigerant chills the air to the point where all the moisture in the air precipitates and is drained away. Refrigeration air dryers are also effective at removing compressor lubrication oil from the air.

Desiccant air dryers circulate air through a container filled with chemical pellets that absorb all the moisture from the air. Silica gel is a common desiccant. The pellets should be replaced regularly. Desiccant air dryers can only be used with air compressors if an aftercooler is installed upstream of the air dryer to cool the air. Both types of air dryer should be drained daily. Refer to your local code before installing an aftercooler or air dryer.

WARNING!

While some types of air compressors can be used in respiration applications, none of the compressors described in this module can be used for respiration or SCUBA (self-contained underwater breathing apparatus) applications. Oil-filled air entering the lungs under pressure is a serious health hazard. Codes prohibit the installation of compressors with oil-lubricated bearings in medical or respiration applications.

Figure 7 ♦ Flexible metal hose used for absorbing vibrations in an air compressor.

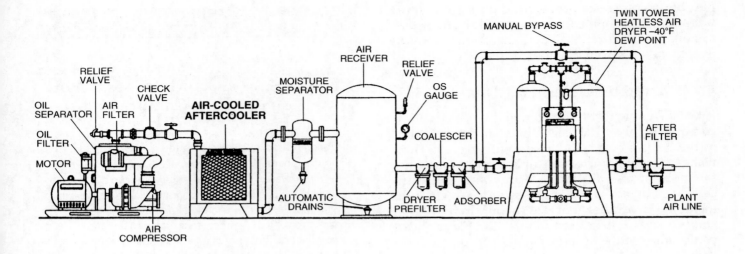
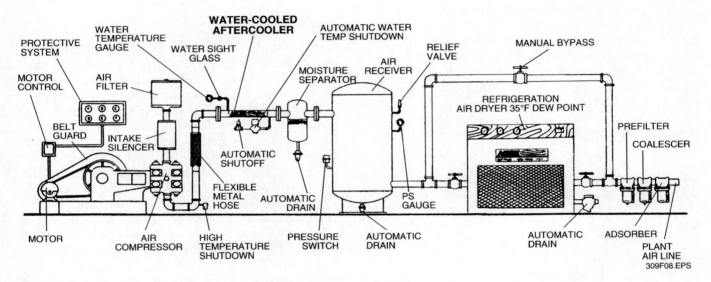

Figure 8 ◆ Typical air- and water-cooled compressed air system installation.

4.3.0 Piping

When the design calls for supply pipe that is 2½ inches in diameter or larger, use standard weight Schedule 40 black iron pipe. Use of galvanized steel is discouraged because flakes from the galvanizing layer can come loose in the air stream. These flakes could damage air and hydraulic components. Some large industrial systems call for welded steel pipe.

For systems that require pipe with a diameter smaller than 2 inches, use Type L copper tubing. When installing this type of pipe in a compressed air system, use either cast brass or wrought copper solder joint fittings. Red brass pipe with threaded cast brass fittings can also be used.

Install piping for compressed air systems using the methods you have already learned. Your local code will specify the correct intervals for pipe hangers and supports. Use only fasteners that are permitted by code. Never use a black iron fitting designed for drainage systems in a compressed air system. The interior of such a fitting is shaped for the flow of liquids, not air.

4.4.0 Controls

Compressed air systems are designed to operate within a specified **control range**, the range between the highest and lowest allowable air pressures. If the system pressure falls below the control range, the machines will not be able to operate properly. If the air pressure is too high, the system and the machines could be damaged. Compressed air systems use a variety of controls to keep the system pressure within the control range. The controls are installed on the air compressor and in the distribution system. The number and types of controls depend on the size and complexity of the system.

4.4.1 Air Compressor Controls

A **stop/start control** turns the air compressor off when the system pressure exceeds the control range. The stop/start control is a simple pressure switch connected to a pressure gauge. Stop/start controls should be used only in compressed air systems that do not cycle frequently. They are intended for light-duty systems.

Throttles control the airflow into an air compressor. The throttle widens and narrows the air compressor's intake valve. Adjusting the valve modulates the output. Throttles are used in centrifugal and rotary compressors.

Constant speed controls are more complex than stop/start and throttling controls. They allow excess air pressure to bleed out of the air compressor when the control range is exceeded. This is called **unloading** the air compressor. A constant speed control allows the motor to run as the air compressor unloads. The system's energy efficiency may drop as a result.

Part-load controls allow air compressors to operate at a percentage of their normal output. For example, a four-step control allows the operator to select an output of 25, 50, 75, or 100 percent.

Many large compressed air systems use more than one compressor. Two types of controls can be used to adjust the pressure in such a system. **Sequencing controls** adjust the overall system pressure by turning individual compressors on or off. Sequencing controls are also called single-master controls. **Network controls** are more complex and efficient. Individual air compressor controls are linked together into a single network controlled by a computer. Network controls allow for more precise control of the system pressure. Network controls are also called multimaster controls.

4.4.2 Distribution System Controls

Shutoff valves cut off the airflow in the supply line in an emergency. Use either ball or gate valves for this purpose. Spring-loaded, plug-style **air throttle valves** are suitable for use as shutoff valves. They can be used as throttles on individual tools as well. Ensure that the pressure rating of the valve is suitable for the application.

High air pressure can damage tools and equipment. **Air pressure regulators** (see *Figure 9*) ensure that the air pressure remains below the high end of the control range. The regulator valve opens, allowing excess pressure to bleed into the regulator body. When the line pressure falls within the control zone again, the valve closes. Regulators are equipped with a pressure gauge. Install an air filter ahead of the regulator to trap particles that could get stuck in the regulator.

Some systems may also require a **lubricator** downstream from the regulator (see *Figure 10*). A lubricator injects a fine oil mist into the supply line. The oil mist lubricates the tools, equipment, and machines powered by compressed air. The manufacturer's instructions will list the types of oil that are safe to use.

The connection between the compressed air system and the tool or machine using the air can

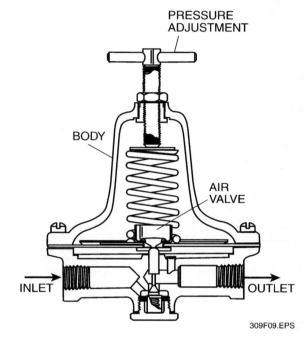

Figure 9 ◆ Cross section of an air pressure regulator.

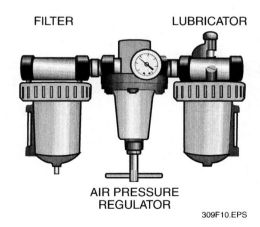

Figure 10 ◆ Filter, pressure regulator, and lubricator installation.

be a location of pressure loss. Pressure is maintained through the use of properly sized **couplings** (see *Figure 11*). A coupling is a snap-fit plug that connects a tool or machine to the compressed air system. A spring-loaded valve prevents airflow until the connection is secure.

5.0.0 ◆ INSTALLING COMPRESSED AIR SYSTEMS

A good compressed air system provides the right amount of high-quality air at the correct pressure. Like other plumbing systems, compressed air systems must be installed correctly to work well. Work closely with the architect, the customer, and the manufacturers. They can give you information about the desired volumes, pressures, and **use factors**.

Install the air compressor, air receiver, filters, and dryers below the tools and machines that they serve. A basement is a good location. Like water supply systems, typical compressed air distribution systems have the following components:

- A main line
- Risers
- Branch lines
- Feeder lines

Locate takeoffs for branch lines and risers where tools and machines will be used. Install feeder lines to connect the branch lines to individual tools and machines. Install valves on all risers, branches, and connections that are installed in anticipation of future system expansion. The takeoffs for all feeder and branch lines should be at the top of the main lines. Install mains and branches so that they slope downward in the direction the air is flowing. Refer to your local code for the proper degree of slope. Use approved hangers and install them at the proper intervals.

Aftercoolers and dryers remove water vapor from air before it enters the distribution system. Another option is to design the distribution system so that it removes the water. Install a **water drip leg** at the end of the main line and in branch lines. Water drip legs are vertical pipes into which water can drain. Install either a manual blowdown valve or an automatic drain valve on each water drip leg. The valve prevents the water drip leg from disrupting the airflow. Install water drip legs at regular intervals along the distribution system. One leg every 50 to 100 feet should be enough.

If the compressed air system is installed correctly, it will require only periodic maintenance. Low pressure is one of the most common problems in compressed air systems. The following are common causes of low air pressure:

- Inadequate sizing of pipes and hoses
- Excessive leakage
- Insufficient air compressor capacity

When inspecting or servicing a compressed air system, consider each of these factors. Ensure that the piping, hoses, and air compressors are all properly sized for the needs of the system. Inspect the system for leaks, which are common at fitting connections and pipe bends. Check to see whether the couplings from the air supply to the tool or machine are the correct size and type. A careful installation will prevent unnecessary service calls in the future.

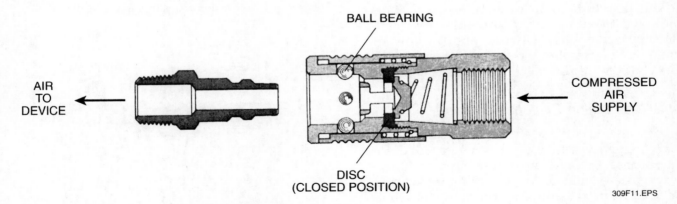

Figure 11 ◆ Cross section of a coupling used in compressed air systems.

Review Questions

Section 4.0.0–5.0.0

1. The most common types of air compressors are _____ and _____.
   a. Rotary; reciprocating
   b. Rotary; centrifugal
   c. Reciprocating; centrifugal
   d. Refrigeration; dessicant

2. Use _____ pipe for systems that require pipe with a diameter larger than 2½ inches.
   a. wrought copper
   b. Schedule 40 black iron
   c. Type L copper
   d. red brass

3. The process of allowing excess air pressure to bleed out of the air compressor when the control range is exceeded is called _____.
   a. throttling
   b. sequencing
   c. part-loading
   d. unloading

4. Spring-loaded, plug-style _____ can be used as shutoff valves.
   a. air pressure regulators
   b. air receivers
   c. air throttle valves
   d. stop/start controls

5. Air pressure regulators ensure that the air pressure remains _____ the high end of the control range.
   a. above
   b. below
   c. equal to
   d. approximate to

6. Like some water supply systems, typical compressed air systems have a main line and branch and feeder lines.
   a. True
   b. False

7. Install mains and branches so that they _____ the direction the air is flowing.
   a. slope downward in
   b. are even with
   c. are above
   d. tilt upward away from

8. When troubleshooting low air pressure, all of the following are possible causes *except*:
   a. too much air leakage
   b. pipes and hoses too small
   c. air compressor capacity too small
   d. too much water vapor

9. Install air compressors _____ the tools or machines they serve.
   a. above
   b. below
   c. near
   d. parallel to

10. To prevent a disruption of the airflow in a compressed air system, install either manual blowdown valves or _____ on water drip legs.
    a. globe valves
    b. temperature and pressure valves
    c. air regulator valves
    d. automatic drain valves

Summary

Compressed air systems provide air for a wide variety of tools, machines, and equipment. In this module, you learned how to size and install large compressed air systems designed to serve an entire building. Compressed air systems operate under the principles of pneumatics. Temperature, pressure, density, and friction are each an important aspect of pneumatics.

Air pressure plays a key role in many plumbing operations. Compressed air pressure is measured in psig. The flow rate of compressed air is measured in cfm. Air compressor inlet capacity is measured in acfm, which measures the actual inlet conditions. Another common measure of air pressure is scfm, which measures air volume at standard conditions.

This module discussed the components of compressed air systems. These systems consist of air compressors, aftercoolers and air dryers, supply piping, and controls. Air compressors condense free air to a specific pressure and distribute it. Most systems use either reciprocating or rotary compressors. Aftercoolers remove a large amount of moisture from compressed air, and refrigeration and desiccant air dryers remove the remaining moisture. Schedule 40 black iron is the most common type of pipe used in a compressed air system. A variety of compressor and distribution controls maintain system pressure within the control range.

Compressed air systems are designed to meet the customer's volume, pressure, and conditioning needs. The skills and techniques required for installing compressed air systems are the same as those required for installing water supply systems.

Air compressors, air receivers, filters, and dryers are usually installed in the basement. The main line, risers, branch lines, and feeder lines are installed according to local code. Water drip legs can be installed to allow water to drain out of the supply pipes. After installation, inspect the system for leaks. If compressed air systems are correctly installed and regularly maintained, they will provide years of reliable service.

Notes

Trade Terms Introduced in This Module

Actual cubic feet per minute: A measure of the air volume at the air compressor inlet under actual temperature and pressure conditions. It is often abbreviated acfm.

Aftercooler: A device that cools air by passing it through a heat exchanger, causing moisture to condense out of the air.

Air dryer: A device that removes all moisture from compressed air.

Air end: A chamber at the end of the rotary screws in a rotary compressor where air is compressed.

Air pressure regulator: A valve that keeps air pressure within the control range.

Air receiver: A storage tank for compressed air.

Air throttle valve: A spring-loaded valve used to control or shut off airflow in a compressed air system.

Boyle's law: A scientific principle stating that air volume increases in direct proportion to a decrease in air pressure. Boyle's law can be expressed as the formula $P_1V_1 = P_2V_2$.

Compressible: Able to be reduced in volume.

Constant speed control: A control that vents excess air pressure from an air compressor.

Control range: The range between the highest and lowest allowable air pressures in a compressed air system.

Coupling: A snap-fit plug connecting a tool or machine to a compressed air system.

Desiccant air dryer: A type of air dryer in which chemical pellets absorb moisture from compressed air.

Hydraulics: The set of physical laws governing the motion of water and other liquids.

Intercooler: A heat exchanger that removes heat from air as it moves through a multistage air compressor system.

Lubricator: A device that injects a fine oil mist into the air supply line, lubricating tools and machines.

Network control: A device that controls multiple compressors by linking them together into a single network controlled by a computer.

Part-load control: A device that allows an air compressor to operate at a given percentage of its normal output.

Pneumatics: The set of physical laws governing the motion of air and other gases.

Reciprocating compressor: A type of air compressor in which air is compressed by a piston.

Refrigeration air dryer: A type of air dryer in which air is cooled as it passes through a heat exchanger cooled by a liquid refrigerant.

Rotary compressor: A type of air compressor in which rotary screws compress the air.

Rotary screw: Helical devices in a rotary compressor that compress air by passing it down the screw path.

Sequencing control: A device that adjusts air pressure in a compressed air system by activating and deactivating individual compressors.

Standard conditions: An air pressure of 14.7 psia, temperature of 68°F, and barometric pressure of 29.92 inches of mercury.

Standard cubic feet per minute: A measure of the air volume at the air compressor inlet under standard conditions. Standard cubic feet per minute is often abbreviated scfm.

Stop/start control: A device that turns an air compressor off when the control range is exceeded.

Throttle: A device that controls airflow into an air compressor by adjusting the compressor's intake valve.

Unloading: The process of relieving excess air pressure from an air compressor.

Use factor: Expressed as a percentage of an average hour of use, the amount of time that a tool or machine will be operated.

Water drip leg: A vertical pipe on an air supply line that acts as a drain for condensed water in the line.

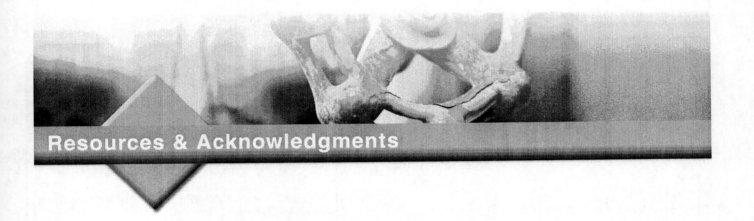

Resources & Acknowledgments

Additional Resources

This module is intended to be a thorough resource for task training. The following reference works are suggested for further study. These are optional materials for continued education rather than for task training.

Compressed Air Systems: A Guidebook on Energy and Cost Savings. 1993. E. M. Talbot. Lilburn, GA: Fairmont Press.

Pneumatic Handbook. 1997. Antony Barber. New York: Elsevier Science.

Pneumatic Systems: Principles and Maintenance. 1996. S. R. Majumdar. New York: McGraw-Hill.

Figure Credits

Atlas Copco, 309F03, 309F04

Campbell Hausfeld, 309F02, 309F06

General Air Products, Inc., 309F05, 309F07

International Association of Plumbing and Mechanical Officials, 2003 *Uniform Plumbing Code™*, Definitions of contamination and pollution

References

Compressed Air Challenge Website, http://www.compressedairchallenge.org, Compressed Air System Fact Sheet #1, "Assessing Compressed Air Needs," http://www.compressedairchallenge.org/conten/library/factsheets/factsh1.pdf, reviewed August 2005.

Compressed Air Challenge Website, http://www.compressedairchallenge.org, Compressed Air System Fact Sheet #4, "Pressure Drop and Controlling System Pressure," http://www.compressedairchallenge.org/content/library/factsheet/factsh4.pdf, reviewed August 2005.

Compressed Air Challenge Website, http://www.compressedairchallenge.org, Compressed Air System Fact Sheet #6, "Compressed Air System Controls," http://www.compressedairchallenge.org/content/library/factsheet/factsh6.pdf, reviewed August 2005.

Dekker Vacuum Technologies Website, http://www.dekkervacuum.com, "Technical Data: Inlet Volume Calculations," pp. 2-015 and 2-016, http://www.dekkervacuum.com/reference/pdf/s2p015n.pdf and /s2p016n.pdf, reviewed August 2005.

Dictionary of Architecture and Construction, Third Edition. 2000. Cyril M. Harris, ed. New York: McGraw-Hill.

Pipefitters Handbook. Forrest R. Lindsey. New York: Industrial Press Inc.

CONTREN® LEARNING SERIES — USER UPDATE

The NCCER makes every effort to keep these textbooks up-to-date and free of technical errors. We appreciate your help in this process. If you have an idea for improving this textbook, or if you find an error, a typographical mistake, or an inaccuracy in NCCER's Contren® textbooks, please write us, using this form or a photocopy. Be sure to include the exact module number, page number, a detailed description, and the correction, if applicable. Your input will be brought to the attention of the Technical Review Committee. Thank you for your assistance.

Instructors – If you found that additional materials were necessary in order to teach this module effectively, please let us know so that we may include them in the Equipment/Materials list in the Annotated Instructor's Guide.

Write: Product Development and Revision
National Center for Construction Education and Research
P.O. Box 141104, Gainesville, FL 32614-1104

Fax: 352-334-0932

E-mail: curriculum@nccer.org

Craft _____ Module Name _____

Copyright Date _____ Module Number _____ Page Number(s) _____

Description
_____
_____
_____
_____

(Optional) Correction
_____
_____
_____

(Optional) Your Name and Address
_____
_____
_____

Glossary of Trade Terms

Actual cubic feet per minute: A measure of the air volume at the air compressor inlet under actual temperature and pressure conditions. It is often abbreviated acfm.

Adhesive: A chemical used to bond thermoset pipes and fittings without fusing them together.

Adsorption: A type of absorption that happens only on the surface of an object.

Aftercooler: A device that cools air by passing it through a heat exchanger, causing moisture to condense out of the air.

Air admittance vent: A valve-type vent that ventilates a stack using air inside a building. The vent opens when exposed to reduced pressure in the vent system, and it closes when the pressure is equalized.

Air dryer: A device that removes all moisture from compressed air.

Air end: A chamber at the end of the rotary screws in a rotary compressor where air is compressed.

Air lock: A pump malfunction caused by air trapped in the pump body, interrupting wastewater flow.

Air pressure regulator: A valve that keeps air pressure within the control range.

Air receiver: A storage tank for compressed air.

Air throttle valve: A spring-loaded valve used to control or shut off airflow in a compressed air system.

Alum: A chemical used for coagulation. It consists of aluminum and sulfur.

Applied mathematics: Any mathematical process used to accomplish a task.

Area: A measure of a surface, expressed in square units.

Atmospheric vacuum breaker: A backflow preventer designed to prevent back siphonage by allowing air pressure to force a silicon disc into its seat and block the flow.

Austenitic stainless steel: A stainless steel alloy that contains nickel and stays strong at very low temperatures. Austenitic steel is the most popular type of steel for corrosive waste pipes.

Back pressure: Excess air pressure in vent piping that can blow trap seals out through fixtures. Back pressure can be caused by the weight of the water column or by downdrafts through the stack vent.

Back siphonage: A form of backflow caused by subatmospheric pressure in the water system, which results in siphoning of water from downstream.

Back vent: An individual vent that is installed directly at the back of a single fixture and connects the fixture drain pipe to the vent stack or the stack vent.

Backflow preventer with intermediate atmospheric vent: A specialty backflow preventer used on low-flow and small supply lines to prevent back siphonage and back pressure. It is a form of double-check valve.

Backflow: An undesirable condition that results when nonpotable liquids enter the potable water supply by reverse flow through a cross-connection.

Backwashing: The cleaning process of a fully automatic water softener.

Beaded end: A raised lip around the rim of borosilicate glass pipe.

Bimetallic thermometer: A thermometer that determines temperature by using the thermal expansion of a coil of metal consisting of two thinner strips of metal.

Biohazard: An organism or chemical that is dangerous to the health of living creatures.

Borosilicate glass: A pipe material made from specially reinforced glass that does not expand or contract much with changes in temperature.

Boyle's law: A scientific principle stating that air volume increases in direct proportion to a decrease in air pressure. Boyle's law can be expressed as the formula $P_1V_1 = P_2V_2$.

Branch interval: The space between two branches connecting with a main stack. The space between branch intervals is usually story height but cannot be less than 8 feet.

Branch vent: A vent that connects fixtures to the vent stack.

Calcium hypochlorite: The solid form of chlorine. It comes in powder and tablet form.

Celsius scale: A centigrade scale used to measure temperature.

Centigrade scale: A scale divided into 100 degrees. Generally used to refer to the metric scale of temperature measure (see Celsius scale).

Glossary of Trade Terms

Centrifugal pump: A device that uses an impeller powered by an electric motor to draw wastewater out of a sump and discharge it into a drain line.

Chlorination: The use of chlorine to disinfect water.

Chlorinator: A device that distributes chlorine in a water supply system.

Circle: A surface consisting of a curve drawn all the way around a point. The curve keeps the same distance from that point.

Circuit vent: A vent that connects a battery of fixtures to the vent stack.

Coagulation: The bonding that occurs between contaminants in the water and chemicals introduced to eliminate them.

Combination waste and vent system: A line that serves as a vent and also carries wastewater. Sovent® systems are an example of a combination waste and vent system.

Common vent: A vent that is shared by the traps of two similar fixtures installed back-to-back at the same height. It is also called a unit vent.

Compressible: Able to be reduced in volume.

Conduction: The transfer of heat energy from a hot object to a cool object.

Conductor: A vertical spout or pipe that drains storm water into a main drain. It is also called a leader.

Constant speed control: A control that vents excess air pressure from an air compressor.

Contamination: An impairment of the quality of the potable water which creates an actual hazard to the public health through poisoning or through the spread of disease by sewage, industrial fluids, or waste. Also defined as High Hazard. (As defined in the *2003 Uniform Plumbing Code*.)

Continuous vent: A vertical continuation of a drain that ventilates a fixture.

Control range: The range between the highest and lowest allowable air pressures in a compressed air system.

Controlled-flow roof drainage system: A roof drainage system that retains storm water for an average of 12 hours before draining it. Controlled-flow systems help prevent overloads in the storm-sewer system during heavy storms.

Conventional roof drainage system: A roof drainage system that drains storm water as soon as it falls.

Corrosive waste: Waste products that contain harsh chemicals that can damage pipe and cause illness, injury, or even death.

Coupling: A snap-fit plug connecting a tool or machine to a compressed air system.

Coupling: Removable metal clamps with plastic seals used to connect borosilicate glass pipe.

Cross-connection: A direct link between the potable water supply and water of questionable quality.

Cube: A rectangular prism in which the lengths of all sides are equal. When used with another form of measurement, such as cubic meter, the term refers to a measure of volume.

Cubic foot: The basic measure of volume in the English system. There are 7.48 gallons in a cubic foot.

Cubic meter: The basic measure of volume in the metric system. There are 1,000 liters in a cubic meter.

Cylinder: A pipe- or tube-shaped space with a circular cross section.

Decimal of a foot: A decimal fraction where the denominator is either 12 or a power of 12.

Demand: The measure of the water requirement for the entire water supply system.

Density: The amount of a liquid, gas, or solid in a space, measured in pounds per cubic foot.

Desiccant air dryer: A type of air dryer in which chemical pellets absorb moisture from compressed air.

Developed length: The length of all piping and fittings from the water supply to a fixture.

Diaphragm pump chlorinator: A type of chlorinator that uses a diaphragm pump to circulate chlorine.

Diatomaceous earth: Soil consisting of the microscopic skeletons of plants that are all locked together.

Disinfection: The destruction of harmful organisms in water.

Distillation: The process of removing impurities from water by boiling the water into steam.

Glossary of Trade Terms

Double-check valve assembly (DCV): A backflow preventer that prevents back pressure and back siphonage through the use of two spring-loaded check valves which seal tightly in the event of backflow. DCVs are larger than dual-check valve backflow preventers and are used for heavy-duty protection. DCVs have test cocks.

Drainage fixture unit (DFU): A measure of the waste discharge of a fixture in gallons per minute. DFUs are calculated from the fixture's rate of discharge, the duration of a single discharge, and the average time between discharges.

Drainage fixture unit: A measure of the discharge in gallons per minute of a fixture, divided by the number of gallons in a cubic foot.

Dual-check valve backflow preventer assembly (DC): A backflow preventer that uses two spring-loaded check valves to prevent back siphonage and back pressure. DCs are smaller than double-check valve assemblies and are used in residential installations. DCs do not have test cocks.

Duplex pump: Two centrifugal pumps and their related equipment installed in parallel in a sump.

Duplex stainless steel: A stainless steel alloy mix of austenitic and ferritic stainless steels that offers very good corrosion resistance.

Ejector: A common alternative name for a sewage pump.

Electrical thermometer: A thermometer that measures temperature by converting heat into electrical resistance or voltage.

English system: One of the standard systems of weights and measures. The other system is the metric system.

Equilibrium: A condition in which all objects in a space have an equal temperature.

Equivalent length: The length of pipe required to create the same amount of friction as a given fitting.

Fahrenheit scale: The scale of temperature measurement in the English system.

Ferritic stainless steel: A stainless steel alloy that is very resistant to rust and has a very low carbon content.

Flat vent: A vent that runs horizontally to a drain. It is also called a horizontal vent.

Float switch: A container filled with a gas or liquid that measures wastewater height in a sump and activates the pump when the float reaches a specified height.

Floc: Another term for precipitate.

Flow rate: The rate of water flow in gallons per minute that a fixture uses when operating.

Friction loss: The partial loss of system pressure due to friction. Friction loss is also called pressure loss.

Friction: The resistance that results from objects rubbing against one another.

Fulcrum: In a lever, the pivot or hinge on which a bar, a pole, or other flat surface is free to move.

Furan: A resin that is highly resistant to solvents. Furans are used to make thermoset pipe.

Fusion coil: A device that uses heat to join thermoplastic pipe.

Gallon: In the English system, the basic measure of liquid volume. There are 7.48 gallons in a cubic foot.

Grains per gallon: A measure of the hardness of water. One grain equals 17.1 ppm of hardness.

Hazard communication: A workplace safety program that informs and educates employees about the risks from chemicals.

Head: The height of a water column, measured in feet. One foot of head is equal to 0.433 pounds per square inch gauge.

High hazard: A classification denoting the potential for an impairment of the quality of the potable water which creates an actual hazard to the public health through poisoning or through the spread of disease by sewage, industrial fluids, or waste. This type of water quality impairment is called contamination.

Hose connection vacuum breaker: A backflow preventer designed to be used outdoors on hose bibbs to prevent back siphonage.

Hydraulic gradient: The maximum degree of allowable fall between a trap weir and the opening of a vent pipe. The total fall should not exceed one pipe diameter from weir to opening.

Hydraulics: The set of physical laws governing the motion of water and other liquids.

Impeller: A set of vanes or blades that draws wastewater from a sump and ejects it into the discharge line.

Trade Terms Introduced in This Module

Inclined plane: A straight and slanted surface.

Indirect or momentum siphonage: The drawing of a water seal out of the trap and into the waste pipe by lower-than-normal pressure. It is the result of discharge into the drain by another fixture in the system.

Individual vent: A vent that connects a single fixture directly to the vent stack. Back vents and continuous vents are types of individual vents.

Injector chlorinator: A type of chlorinator that uses an injector to circulate chlorine.

In-line vacuum breaker: A specialty backflow preventer that uses a disc to block flow when subjected to back siphonage. It works the same as an atmospheric vacuum breaker.

Intercooler: A heat exchanger that removes heat from air as it moves through a multistage air compressor system.

Ion exchange: A technique used in water softeners to remove hardness by neutralizing the electrical charge of the mineral atoms.

Ion: An atom with a positive or negative electrical charge.

Isosceles triangle: A triangle in which two of the sides are of equal length.

Kelvin scale: The scale of temperature measurement in the metric system.

Laminar flow: The parallel flow pattern of a liquid that is flowing slowly. Also called streamline flow or viscous flow.

Lever: A simple machine consisting of a bar, a pole, or other flat surface that is free to pivot around a fulcrum.

Lift station: A popular name for a sewage removal system. The term is also used to refer to a pumping station on a main sewer line.

Liquid thermometer: A thermometer that measures temperature through the expansion of a fluid, such as mercury or alcohol.

Liter: In the metric system, the basic measure of liquid volume. There are 1,000 liters in a cubic meter.

Loop vent: A vent that connects a battery of fixtures with the stack vent at a point above the waste stack.

Low hazard: A classification denoting the potential for an impairment in the quality of the potable water to a degree which does not create a hazard to the public health but which does adversely and unreasonably affect the aesthetic qualities of such potable water for domestic use. This type of water quality impairment is called pollution.

Lubricator: A device that injects a fine oil mist into the air supply line, lubricating tools and machines.

Manufactured air gap: An air gap that can be installed on a reduced-pressure zone principle backflow preventer to prevent back pressure and back siphonage.

Martensitic stainless steel: A stainless steel alloy that is very strong but is mostly used to handle mild corrosive wastes.

Mercury float switch: A float switch filled with mercury. It activates the pump when it reaches a specified height in a basin.

Metric system: A system of measurement in which multiples and fractions of the basic units of measure are expressed as powers of 10. Also called the SI system.

Network control: A device that controls multiple compressors by linking them together into a single network controlled by a computer.

NFPA diamond: A label that uses color-coding and numbers to indicate the properties of chemicals. The National Fire Protection Association developed the diamond symbol.

Osmosis: The unequal back-and-forth flow of two different solutions as they seek equilibrium.

Oxidizing agent: A chemical made up largely of oxygen used for oxidation.

Part-load control: A device that allows an air compressor to operate at a given percentage of its normal output.

Pasteurization: Heating water to kill harmful organisms in it.

Percolate: To seep into the soil and become absorbed by it.

Permeability: A measure of the soil's ability to absorb water that percolates into it.

Pneumatic ejector: A pump that uses compressed air to force wastewater from a sump into the drain line.

Trade Terms Introduced in This Module

Pneumatics: The set of physical laws governing the motion of air and other gases.

Point–of–entry unit: A water conditioner that is located near the entry point of a water supply.

Point–of–use unit: A water conditioner that is located at an individual fixture.

Pollution: An impairment of the quality of the potable water to a degree which does not create a hazard to the public health but which does adversely and unreasonably affect the aesthetic qualities of the potable water for domestic use. Also defined as Low Hazard. (As defined in the 2003 Uniform Plumbing Code.)

Ponding: Collecting of storm water on the ground.

Pounds per square inch: In the English system, the basic measure of pressure. Pounds per square inch (psi) is measured in pounds per square inch absolute (psia) and pounds per square inch gauge (psig).

Precipitate: A particle created by coagulation that settles out of the water.

Precipitation: The process of removing contaminants from water by coagulation.

Pressure drop: In a water supply system, the difference between the pressure at the inlet and the pressure at the farthest outlet.

Pressure float switch: An adjustable float switch that can either be allowed to rise with wastewater or be clipped to the sump wall.

Pressure: The force applied to the walls of a container by the liquid or gas inside.

Pressure-type vacuum breaker: A backflow preventer installed in a water line that uses a spring-loaded check valve and a spring-loaded air inlet valve to prevent back siphonage.

Prism: A volume in which two parallel rectangles, squares, or right triangles are connected by rectangles.

Probe switch: An electrical switch used in pneumatic ejectors that closes on contact with wastewater.

Pulley: A simple machine consisting of a rope wrapped around a wheel.

Reciprocating compressor: A type of air compressor in which air is compressed by a piston.

Rectangle: A four-sided surface in which all corners are right angles.

Reduced-pressure zone principle backflow preventer assembly: A backflow preventer that uses two spring-loaded check valves plus a hydraulic, spring-loaded pressure differential relief valve to prevent back pressure and back siphonage.

Refrigeration air dryer: A type of air dryer in which air is cooled as it passes through a heat exchanger cooled by a liquid refrigerant.

Relief vent: A vent that increases circulation between the drain and vent stacks, thereby preventing back pressure by allowing excess air pressure to escape.

Retention pond: A large man-made basin into which storm water runoff drains until sediment and contaminants have settled out.

Retention tank: An enclosed container that stores storm water runoff until sediment and contaminants have settled out.

Re-vent: To install an individual vent in a fixture group.

Reverse flow pump: A duplex pump designed to discharge solid waste without allowing it to come into contact with the pump's impellers.

Right triangle: A three-sided surface with one angle that equals 90 degrees.

Rotary compressor: A type of air compressor in which rotary screws compress the air.

Rotary screw: Helical devices in a rotary compressor that compress air by passing it down the screw path.

Screw: A simple machine consisting of an inclined plane wrapped around a cylinder.

Scupper: An opening on a roof or parapet that allows excess water to empty into the surface drain system.

Secondary drain: A drain system that acts as a backup to the primary drain system. It is also called an emergency drain.

Self-siphonage: A condition whereby lower-than-normal pressure in a drain pipe draws the water seal out of a fixture trap and into the drain.

Sequencing control: A device that adjusts air pressure in a compressed air system by activating and deactivating individual compressors.

Trade Terms Introduced in This Module

Sewage pump: A device that draws wastewater from a sump and pumps it to the sewer line. It may be either a centrifugal pump or a pneumatic ejector.

Sewage removal system: An installation consisting of a sump, pump, and related controls that stores sewage wastewater from sub-drains and lifts it into the main drain line.

SI system: An abbreviation for Système International d'Unités, or International System of Units. The formal name of the metric system.

Simple machine: A device that performs work in a single action. There are six types of simple machines: the inclined plane, the lever, the pulley, the wedge, the screw, and the wheel and axle.

Sodium hypochlorite: The liquid form of chlorine, commonly found in laundry bleach.

Solenoid valve: An electronically operated plunger used to divert the flow of water in a water conditioner.

Sovent® system: A combination waste and vent system used in high-rise buildings that eliminates the need for a separate vent stack. The system uses aerators on each floor to mix waste with air in the stack and a de-aerator at the base of the stack to separate the mixture.

Square foot: In the English system, the basic measure of area. There are 144 square inches in one square foot.

Square meter: In the metric system, the basic measure of area. There are 10,000 square centimeters in a square meter.

Square: A rectangle in which all four sides are equal lengths. When used with another form of measurement, such as square meter, the term refers to a measure of area.

Stack vent: An extension of the main soil and waste stack above the highest horizontal drain. It allows air to enter and exit the plumbing system through the building roof and is a terminal for other vent pipes.

Standard conditions: An air pressure of 14.7 psia, temperature of 68°F, and barometric pressure of 29.92 inches of mercury.

Standard cubic feet per minute: A measure of the air volume at the air compressor inlet under standard conditions. Standard cubic feet per minute is often abbreviated scfm.

Stop/start control: A device that turns an air compressor off when the control range is exceeded.

Storm water removal system: An installation consisting of a sump, pump, and related controls that stores storm water runoff from a sub-drain and lifts it into the main drain line.

Stroke: A single operating cycle of a diaphragm pump.

Sub-drain: A building drain located below the sewer line.

Sump pump: A common name for a pump used in a storm water removal system.

Sump: A container that collects and stores wastewater from a sub-drain until it can be pumped into the main drain line.

Tablet chlorinator: A type of chlorinator in which water is pumped through a container of calcium hypochlorite tablets.

Temperature: A measure of relative heat as measured by a scale.

Test cock: A valve in a backflow preventer that permits the testing of individual pressure zones.

Thermal expansion: The expansion of materials in all three dimensions when heated.

Thermometer: A tool used to measure temperature.

Thermoplastic: A type of plastic pipe that can be softened and reshaped when it is exposed to heat.

Thermoset: A type of plastic pipe that is hardened into shape when manufactured.

Throttle: A device that controls airflow into an air compressor by adjusting the compressor's intake valve.

Transient flow: The erratic flow pattern that occurs when a liquid's flow changes from a laminar flow to a turbulent flow pattern.

Turbidity unit: A measurement of the percentage of light that can pass through a sample of water.

Turbulent flow: The random flow pattern of fast-moving water or water moving along a rough surface.

Ultraviolet light: Type of disinfection in which a special lamp heats water as it flows through a chamber.

Uniform Laboratory Hazard Signage: Signs placed on laboratory walls that use pictographs to indicate hazards.

Trade Terms Introduced in This Module

Unloading: The process of relieving excess air pressure from an air compressor.

Use factor: Expressed as a percentage of an average hour of use, the amount of time that a tool or machine will be operated.

Vent stack: A stack that serves as a building's central vent, to which other vents may be connected. The vent stack may connect to the stack vent or have its own roof opening. It is also called the main vent.

Vent terminal: The point at which a vent terminates. For a fixture vent, it is the point where it connects to the vent stack or stack vent. For a stack vent, it is the point where the vent pipe ends above the roof.

Vent: A pipe in a DWV system that allows air to circulate in the drainage system, thereby maintaining equalized pressure throughout.

Viscosity: The measure of a liquid's resistance to flow.

Volume: A measure of a total amount of space, measured in cubic units.

Water conditioner: Device used to remove harmful organisms and materials from water.

Water drip leg: A vertical pipe on an air supply line that acts as a drain for condensed water in the line.

Water Supply Fixture Unit (WSFU): The measure of a fixture's load, depending on water quantity, temperature, and fixture type.

Wedge: A simple machine consisting of two back-to-back inclined planes.

Wet vent: A vent pipe that also carries liquid waste from fixtures with low flow rates. It is most frequently used in residential bathrooms.

Wheel and axle: A simple machine consisting of a circle attached to a central shaft that spins with the wheel.

Work: A measure of the force required to move an object a specified distance.

Yoke vent: A relief vent that connects the soil stack with the vent stack.

Plumbing Level Three

Index

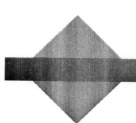

Index

A
ABS. *See* Acrylonitrile butadiene styrene
Absorption, 3.13
acfm. *See* Actual cubic feet per minute
Acid, battery, 8.3
Acids and acidity
  acid dilution and neutralization sumps, 3.19, 8.14, 8.15, 8.18
  causes and effects on pipe, 3.18, 3.21
  corrosive-resistant pipe, 8.2, 8.6
  filters, 3.9, 3.21
  lime to treat, 3.11
  soda ash to treat, 3.14, 3.19, 3.21
  test kits, 3.19, 3.21
Acrylonitrile butadiene styrene (ABS), 6.22
Activated carbon (AC), 3.13–3.14, 3.19, 3.21
Actual cubic feet per minute (acfm), 9.3, 9.13
Adhesive, epoxy, 8.5, 8.12, 8.19
Adsorption, 3.13, 3.22
Aerator, 3.11
Aftercooler, 9.7–9.8, 9.10, 9.12, 9.13
Agriculture
  RPZ backflow preventers on farms, 4.11
  runoff, 4.13, 6.12
Air
  basic principles, 9.3–9.4
  compressed. *See* Compressed air
  in the DWV system, 5.2, 5.12–5.13
  in the pump. *See* Air lock
  in the water system, 4.2
Air end, 9.5, 9.13
Air gap, 4.6, 4.7–4.8, 4.11, 4.12, 4.18
Air lock, 7.14, 7.15, 7.19
Alarms
  high-water sump, 7.9
  power-off, 7.9, 7.13
Alkalines and alkalinity, 3.3, 3.7, 3.8, 3.21, 8.2
Alternator, pump, 7.4
Alum, 3.14, 3.19, 3.22
Aluminum, 1.17, 3.14, 7.10, 8.4
American Society for Testing and Materials International (ASTM), 2.7
American Society of Mechanical Engineers (ASME), 9.5
American Society of Sanitary Engineering (ASSE), 4.3, 5.8
American Water Works Association (AWWA), 2.7, 4.3, 4.15
Ammonia, 3.11, 8.4
Apartment buildings and condominiums, 4.10, 5.10, 6.8

Appliances
  backflow preventers, 4.6, 4.7
  damage by hard water, 3.9
  expansion and contraction, 1.18
Area
  definition and use of term, 1.2, 1.28
  measuring, 1.5–1.10, 1.32
  and rainfall weight/volume calculations, 6.13, 6.18
  units, 1.3
Arrester, water hammer, 1.20, 9.3
Asbestos, 8.8
ASME. *See* American Society of Mechanical Engineers
ASSE. *See* American Society of Sanitary Engineering
ASTM. *See* American Society for Testing and Materials International
Atmospheric vacuum breaker (AVB), 4.6, 4.8–4.9, 4.18
Atomic structure, 3.11
Auger, sewer, 8.13
Auto maintenance facilities, 8.3
Auto painting and detailing shops, 8.3
AWWA. *See* American Water Works Association
Axle, 1.23, 1.26

B
Backflow
  and cross-connections, 4.2, 4.17
  definition, 4.2, 4.17, 4.18
  EPA tracking, 4.4
  pneumatic ejector, 7.6
  pump, 7.10
Backflow preventers
  air gap, 4.6, 4.7–4.8, 4.11, 4.12, 4.18
  atmospheric vacuum breaker, 4.6, 4.8–4.9, 4.18
  bypass prohibited, 4.11
  containment and isolation, 4.6, 4.7
  double-check valve assembly, 4.6, 4.10–4.11, 4.18
  dual-check valve assembly, 4.6, 4.10, 4.18
  function, 4.2, 4.17
  in-line vacuum breaker, 4.14, 4.18
  pressure-type vacuum breaker, 4.6, 4.9–4.10, 4.18
  reduced-pressure zone principle, 4.11–4.13, 4.18
  selection, 4.14, 4.17
  specialty, 4.13–4.14
  spill-proof, 4.6
  standards, 4.3
  testing, 4.14–4.15, 4.17
Backflow preventer tester, 4.14, 4.15, 4.17

Backhoes, 1.23
Backwashing, 3.11, 3.22
Bacteria and viruses
 diseases caused by, 3.2, 4.13
 Escherichia coliform (E. coli), 4.13
 iron-eating, 3.18
 Legionella, 4.13
 removal. *See* Disinfection; Distillation; Filters
 spread by cross-contamination, 4.13
 sulfate-reducing, 3.19
 sulfur, 3.19
 temperature which kills, 3.6
Bar, wet, 5.13
Barium, 8.4
Basin, settling or retention
 in municipal water treatment, 3.11
 in precipitation system, 3.14, 3.15
 sewage, 7.2
 storm water, 6.22, 7.12, 7.13
Bathroom group
 DFU, 6.3, 6.4, 7.8
 vent sizing, 6.8, 6.9
 WSFU, 2.8
Bathtub
 air gap as backflow preventer, 4.7
 demand and flow pressure, 2.7
 DFU, 6.3, 6.4, 7.8
 vents for, 5.7, 5.11
 WSFU, 2.8, 2.9
Benzene, 3.17
Beverage machines, 4.6, 4.7, 6.11
Bicarbonates, 3.17
Bidet
 demand and flow pressure, 2.7
 DFU, 6.3, 7.8
 vents for, 5.11
 WSFU, 2.8
Biohazards, 8.15, 8.16, 8.19
Bit, drill, 1.25
Bleach, 3.3, 3.13, 3.21, 8.2
Blueprints. *See* Drawings
Boiler feed, residential, 4.13
Books and magazines, 5.8
Bottle, hose-attached garden spray, 4.10
Bottled water, 3.3, 4.13
Boyle's law, 9.4, 9.13
Branch, horizontal, 6.4, 6.5, 6.6, 6.11, 6.16, 9.10
Branch interval, 5.13, 5.17, 6.4
Brass, 3.18, 6.22
Brine, 3.11, 3.12
British Imperial system, 1.2
Brownian motion, 1.15
Burns, 3.5, 9.2

C
Calcium, 3.11, 3.14, 8.4
Calcium carbonate, 3.14
Calcium hypochlorite, 3.3, 3.5, 3.22
Calcium salts, 3.2, 3.17, 3.21. *See also* Softening of water
Calculator, square root key, 1.8, 1.9
Calibration, backflow preventer test equipment, 4.15
Canadian Standards Association (CAN/CSA), 2.7
Cancer hazards, 8.15, 8.16
CAN/CSA. *See* Canadian Standards Association
Cap, at pipe end, 3.2, 6.4
Carbon. *See* Activated carbon
Carbonated beverage machines, 4.6, 4.7, 6.11

Carbonation systems, 4.10
Carbon dioxide, 3.18, 4.10
Car wash, 4.4, 4.11
Cast Iron Soil Pipe Institute, 8.6
Catch basin, 1.10
Caulking, 8.12
Cellulose acetate, 3.15
Celsius scale, 1.15, 1.16, 1.19, 1.28
Cement, solvent, 8.8, 8.9, 8.11, 8.12
Centigrade scale, 1.16, 1.28
Centrifugal air compressors, 9.5
Centrifugal pumps, 7.2–7.4, 7.10, 7.11, 7.13, 7.19
Charlotte Pipe, 6.2
Chemicals. *See* Hazardous materials
Chisel, as a wedge, 1.25
Chlorides, 3.17
Chlorinated polyvinyl chloride (CPVC), 7.10, 7.12, 8.4
Chlorinator
 definition, 3.22
 diaphragm pump, 3.3–3.4, 3.22
 injector, 3.4–3.5, 3.22
 tablet, 3.5, 3.22
Chlorine
 disinfection properties, 3.3, 3.21
 to flush ion exchange system, 3.12–3.13
 to flush the water system, 3.2, 3.3
 in municipal water treatment, 3.11
 removal, 3.13, 3.19
 to remove iron-eating bacteria, 3.18
 toxicity, 3.5, 3.12
Circle, 1.8–1.9, 1.11, 1.28
Circumference, 1.9
Clamp, riser, 8.8, 8.9
Cleanout, 1.18, 5.7, 5.10, 8.13
Clean Water Act, 6.15
Climate control systems, 9.2
Coagulation, 3.14, 3.22
Coffee machines, 4.7, 6.11
Coil, fusion, 8.9, 8.10, 8.11
Cold conditions. *See* Freezing conditions
Collar, flashing, 5.5, 5.6
Combined sewer overflow (CSO), 6.13, 6.15
Commercial buildings, 5.11
Communication, hazard, 8.14–8.16, 8.18, 8.19
Compressed air
 basic principles, 9.3–9.4, 9.12
 system components, 9.5–9.10, 9.12
 system installation, 9.10, 9.12
 working safety with, 9.2
Compressibility, 9.3, 9.13
Compressor, 7.9, 9.5–9.7, 9.9, 9.10, 9.12
Concrete, 6.22, 7.6, 7.7, 7.12
Condensation, 5.6, 6.6
Conditioners, water, 3.2
Conduction, 1.14, 1.28
Conductor (leader), 6.13, 6.16, 6.18, 6.24
Confined spaces, 7.15
Conservation measures, 1.12, 2.7
Construction projects, 8.3, 9.2
Containers
 volume calculations, 1.12–1.13
 weight considerations, 1.12
Contamination
 definition, 3.2, 3.22
 during plumbing installation, 3.10
 prevention or treatment. *See* Disinfection; Distillation;
  Filters; Precipitation systems

sulfur-based, 3.19
  through back siphonage. *See* Siphonage, back
Contraction. *See* Thermal expansion and contraction
Controls
  air compressor, 9.8–9.9, 9.13
  air distribution system, 9.9–9.10
  pump, 7.7, 7.9, 7.13
Conversions
  air volume, 9.4
  energy, 1.15
  mathematical, 1.4–1.5, 1.7, 1.12, 1.30–1.31
  rainfall, 6.13–6.15
  temperature, 1.16
Copper
  flashing, 8.8
  pipe, 2.13, 3.18, 6.22
  thermal expansion, 1.17, 1.18
  tubing, 2.13, 9.8
Corrosion, 3.14, 3.21, 7.14, 8.2, 8.18
Couplings
  air compressor, 9.10, 9.13
  definition, 8.3, 8.19
  for glass pipe, 8.3, 8.4, 8.7, 8.9
  for plastic pipe, 8.4, 8.9, 8.10, 8.11
  for silicon cast-iron pipe, 8.5, 8.12, 8.13
CPVC. *See* Chlorinated polyvinyl chloride
Cranes, 1.23
Crawl space, 3.9, 3.10
Crimping tool, 8.14
Cross, sanitary, 5.10
Cross-connection, 4.2, 4.7, 4.13, 4.18
CSO. *See* Combined sewer overflow
Cube, 1.11, 1.28
Cubic feet
  use of term, 1.10, 1.28, 9.3
  weight of water, 1.12, 6.13
Cubic meter, 1.10, 1.28
Cuspidor, dental, 6.3, 7.8
Cutter, glass pipe, 8.3, 8.4
Cylinder, 1.10, 1.11–1.12, 1.13, 1.28, 1.32, 7.12

D

Dairies, 4.11
Darkroom, 4.10
DC. *See* Dual-check valve backflow preventer assembly
DCV. *See* Double-check valve assemblies
Decimals of a foot, 1.6, 1.7, 1.28
Degreaser, 4.9
Demand
  calculation, 2.8–2.10
  definition and use of term, 2.2, 2.17
  intermittent or continuous, 2.10
  peak, 2.7, 2.10
Density, 2.2–2.3, 2.17
Dentist facilities, 6.3
Detergents, 8.2
Developed length, 2.10, 2.17
DFU. *See* Drainage fixture unit
Diameter, 1.9, 1.12, 1.13. *See also* Sizing
Diaphragming, 6.6
Diatomaceous earth, 3.11, 3.19, 3.22
Dilution, 8.14
Discoloration, 3.18, 3.19
Disease. *See* Bacteria and viruses
Dishwashers
  backflow preventers, 4.7, 4.8
  demand and flow pressure, 2.7

DFU, 6.3, 7.8
vents for, 6.11
WSFU, 2.8
Disinfection, 3.2–3.8, 3.11, 3.21, 3.22
Distillation, 3.16, 3.22
Distillation systems, 3.16–3.17, 3.21
Door, as a lever, 1.24
Double-check valve assemblies (DCV), 4.6, 4.10–4.11, 4.18
Downdraft, 5.2
Drain, waste, and vent pipes (DWV)
  air allowed to exit. *See* Vents
  air pressure equilibrium within, 5.2, 5.12–5.13
  calculating force on test plugs, 1.22–1.23
  combination waste and vent system, 5.13, 5.17, 6.11
  grade or slope, 5.5–5.7
  overview, 6.23
  sizing, 6.2–6.11
  in tall or high-rise buildings, 5.13, 5.14
  volume measurement, 1.10–1.13
Drainage fixture unit (DFU)
  definition, 6.24, 7.19
  sizing, 6.2–6.5, 6.7, 6.10, 6.11
  for specific fixtures, 6.3, 7.8
  in sump design, 7.6–7.7, 7.8
Drainage systems
  above-grade and below-grade, 6.22, 8.8–8.9
  roof, 6.13, 6.15–6.22
  sewer, 6.5, 7.13
  storm, 6.13–6.22
Drain cleaner, 8.2
Drains
  air compressor, 9.5
  below sewer lines. *See* Pumps, sewage; Pumps, sump
  curb, 7.13
  diameter and vent sizing, 6.8
  floor, 6.3, 6.5, 7.8, 8.8
  ion exchange system, 3.12
  roof, 6.16, 6.18, 6.19, 6.20, 6.21
  secondary, 6.21, 6.24
  sizing, 6.4–6.6, 7.7
  sub-, 7.2, 7.12, 7.16, 7.19
Drawings
  blueprints, 2.6
  isometric sketch, 2.11
  pipe installation design, 8.8
Drinking fountain
  air gap as backflow preventer, 4.7
  demand and flow pressure, 2.7
  DFU, 6.3, 7.8
  vents for, 5.11
  WSFU, 2.8
Drinking water. *See* Water supply, potable
Driveways, 1.19, 7.12
Dryer, air, 9.5, 9.7–9.8, 9.10, 9.12, 9.13
Dual-check valve backflow preventer assembly (DC), 4.6, 4.10, 4.18
DWV. *See* Drain, waste, and vent pipes

E

Ejector, pneumatic
  capability standards, 7.16
  definition, 7.19
  overview, 7.2, 7.4, 7.6
  probe switches in, 7.9, 7.14
  vents for, 5.14, 7.14
Elbow, 2.4, 8.11
Electrical circuit, motor, 7.15

Electrical hazards, 7.14, 8.15, 8.16
Electrical problems, troubleshooting, 7.14, 7.15
Electric eye, 3.7
Electrodes, 7.9, 7.14
Electrolysis, 8.2
Emphysema, 3.5
End of pipe
   beaded, 8.3, 8.7, 8.19
   cap, 3.2, 6.4
Energy of heat transfer, 1.14
Energy-saving measures, 1.12, 2.7
Engineering, sanitary, 7.16
English system, 1.2, 1.3, 1.19, 1.28, 1.30–1.31
EPA. See U.S. Environmental Protection Agency
Epoxy, 8.5, 8.12
Equilibrium, 1.14, 1.28, 3.15, 5.2
Equivalent length, 2.3, 2.4, 2.14, 2.17
Escherichia coliform (E. coli), 4.13
Expansion. See Thermal expansion and contraction
Expansion coefficient, 1.18
Explosion, 2.2, 5.6, 6.6, 8.12

F
Fahrenheit scale, 1.15, 1.16, 1.19, 1.28
Farms, poultry and dairy, 4.11
Faucet, 4.6, 4.7
FDA. See U.S. Food and Drug Administration
Feeder
   chemical, 3.14, 3.15
   line, on compressed air systems, 9.10
Fiberglass, 7.6, 7.12, 8.5, 8.8
Filters
   air, 9.5, 9.9
   bacterial inhibitors in, 3.14
   carbon, 3.13–3.14, 3.18, 3.19, 3.21
   cartridge, 3.9
   charcoal, 3.3, 3.9
   diatomaceous earth, 3.11, 3.19, 3.22
   mechanical (microfilters), 3.14, 3.21
   membrane, 3.9, 3.15–3.16
   neutralizing, 3.14, 3.21
   order of multiple, 3.14
   oxidizing, 3.14, 3.18, 3.21
   sand, 3.9, 3.11, 3.14, 3.18, 3.19, 3.21
Filtration, 3.2, 3.9–3.11, 3.13–3.17
Fire control system, 2.10, 4.6, 4.7
Fittings
   bell and spigot, 8.12
   expansion and contraction, 1.18
   friction loss due to, 2.3, 2.4
   joining with pipe. See Couplings
   silicon cast-iron, 8.5
   threaded, 8.11
Fixtures
   backflow preventers, 4.6, 4.7, 4.8
   calculation of drainage fixture units, 6.2–6.3, 7.7
   demand and flow pressure, 2.6–2.8
   in developed length, 2.10
   discharge, 6.2, 7.6. See also Drainage fixture unit
   expansion and contraction, 1.18
   installation for future access, 2.10
   restaurant, 4.7
   vent for floor-mounted, 5.6
   vents for back-to-back, 5.10, 5.12, 6.8, 6.9, 6.10
   water supply fixture unit, 1.20, 1.21, 2.8–2.9
Flammable chemicals, 8.15, 8.16
Flashing, 5.5, 5.6, 6.21, 8.8
Flavor. See Taste, foul

Floc, 3.14, 3.22
Flood control, 7.13
Floor, 1.5, 1.10, 2.6, 8.8
Flow
   fluid mechanics, 2.3, 6.6
   peak, 7.4, 7.12. See also Rainfall, maximum expected
   reverse, 7.4, 7.5, 7.15
   and system design, 2.6
   types, 2.3, 2.17
   vertical vs. horizontal, 6.6
Flow rate
   controlled-flow roof drainage, 6.19, 6.20
   definition, 2.17
   effects of friction, 2.3, 2.4
   for fixtures, 2.6–2.8
   and fluid mechanics, 2.3
   for horizontal pipes, 6.6
   in sizing calculation, 2.12
   storm water sump, 7.12
   for vertical stack, 6.6
Flowserve, 8.5
Fluoride, 3.9, 3.13
Flushing the water system, 3.2–3.3, 3.10
Flushometer, 2.7, 2.8
Force
   and pressure, 1.19–1.23
   relationship with work, 1.23
   on test plugs, 1.22–1.23
Foundation, slab-on-grade, 3.9, 3.10
Fractions, reduction, 1.6
Freezers, 1.15
Freezing conditions
   backflow preventer protection, 4.7
   borosilicate glass pipe for liquids, 8.3
   effects on water, 2.2
   joining silicon cast-iron pipe, 8.12
   pipe protection, 1.18–1.19
   vent extension to prevent frost, 6.6
   vent system, 5.5, 5.6
Friction, 2.3, 2.17
Friction loss, 2.3–2.4, 2.12–2.14, 2.17
Fulcrum, 1.24, 1.28
Furan, 8.5, 8.19
Fusion method of pipe joining, 8.9–8.11

G
Gallon
   definition and use of term, 1.10, 1.28
   rainfall conversions, 6.13, 6.15
   weight of water, 1.12
Garage, unattached, 6.8
Gases
   hazardous, 3.5, 6.6, 7.6, 8.2
   medical, 5.3
   sewer, 5.6, 6.6, 7.6
   water vapor, 9.7, 9.10, 9.12
Gastight seal, 8.14
Glass
   borosilicate pipe, 8.3, 8.4, 8.7–8.9, 8.19
   fiberglass, 7.6, 7.12, 8.5, 8.8
   wool, 8.8
Government work, 1.4
Grade, 1.24, 5.5–5.7
Grains per gallon, 3.17, 3.18, 3.22
Graphite, 8.5
Gun, nail, 9.3
Gutter, street, 7.12

H

Hammer (tool), 1.24, 9.3
Hammer (water), 1.20, 1.21, 9.3
Handwheel, 1.26
Hanger
    compressed air system lines, 9.10
    pipe, 1.18, 8.8, 8.12, 8.13, 8.18, 9.8
Hard water, 3.9, 3.11–3.13, 3.16, 3.17–3.18, 3.21
Hazardous materials
    biohazards, 8.15, 8.16, 8.19
    communication, 8.14–8.16, 8.18, 8.19
    corrosive waste, 8.2–8.3, 8.19. *See also* Pipe, corrosive-resistant
    gases, 3.5, 6.6, 7.6, 8.2
    household, 8.2
    NFPA diamond, 8.14–8.15, 8.16, 8.19
    in storm water runoff, 6.13, 7.10, 7.14
HazCom plan, 8.14, 8.16
Head, 1.20–1.22, 1.28
Head loss, static, 2.12
Hearing protection, 9.2
Heat, 1.14, 1.15
Heat exchanger, 3.6, 4.9, 9.6, 9.7
Heating, ventilating, and air conditioning system (HVAC), 5.5
Heating element, 3.16, 3.17, 8.9, 8.10, 8.11
High Hazard classification, 4.5, 4.6, 4.11, 4.14, 4.18
High-rise buildings, 2.7, 5.13, 5.14
Hinge, as a fulcrum, 1.24
History
    compressed air and water tools, 9.3
    corrosive chemicals, 8.2
    discovery of air pressure, 5.4
    metric system, 1.3
    pasteurization, 3.6
    pipe, 2.11, 6.2, 6.21
    pipe sizing, 2.11
    public health and drinking water, 3.2
    pump, 7.3, 7.11
    sanitary engineering, 7.16
    Sovent® vent systems, 5.14
    standardized measures, 1.2
    temperature scales, 1.19
    wheel, 1.26, 7.3
Hoses
    air compressor, 9.7, 9.10
    pneumatic tool, 9.2
    vacuum breaker, 4.6, 4.18
Hospitals, 4.13, 4.14, 5.3
Hotels, 5.10
Hot tub, 4.10
Housing subdivisions, 7.4
HVAC. *See* Heating, ventilating, and air conditioning system
Hydrant, wall, 4.6
Hydraulic gradient (slope), 5.6, 5.7, 5.12, 5.17, 6.5, 6.16
Hydraulics, 9.3, 9.13
Hydrochloric acid, 3.18
Hydrogen sulfide, 3.9, 3.19

I

IAPMO. *See* International Association of Plumbing & Mechanical Officials
ICC. *See* International Code Council
Icemaker, 4.7, 6.11
Illness, 3.5, 3.12. *See also* Bacteria and viruses; Public health issues
Impeller, 7.3, 7.4, 7.14, 7.15, 7.19
Information on the Internet, 5.8, 6.13, 8.15
Insulation, 1.19, 4.7, 9.3
Interceptor, grease and oil, 1.10, 1.12, 6.11
Intercooler, 9.6, 9.13
International Association of Plumbing & Mechanical Officials (IAPMO), 5.8
International Bureau of Weights and Measures, 1.2
International Code Council (ICC), 5.8
International Plumbing Code® (IPC)
    air admittance vents, 5.13
    approved pipe materials, 6.22
    backflow preventer classification, 4.6
    centrifugal pumps, 7.4
    flow pressure standards, 2.6
    metric standards, 1.3, 1.4
    secondary drains, 6.21
    sewage and storm water removal systems, 7.16
    sizing method, 2.11
    standards organizations in, 2.7
    wet vents, 5.11
International System of Units. *See* Metric system
Ion, 3.11, 3.22
Ion exchange systems, 3.11–3.13, 3.21, 3.22
IPC. *See International Plumbing Code®*
Iron
    cast
        pipe, 6.2, 6.21, 6.22, 8.12–8.13
        sumps, 7.12
        thermal expansion, 1.17
    chlorine to remove, 3.3
    discoloration by, 3.18, 3.21
    filter to remove, 3.9
    oxidation to remove, 3.14, 3.21
    Schedule 40 black, 9.8, 9.12
Iron salts, 3.17
Irrigation system, 2.10, 4.9

J

Joints
    glass pipe, 8.3, 8.4, 8.7–8.9, 8.18
    leak prevention, 6.6
    plastic pipe, 8.9–8.12, 8.18
    silicon cast-iron pipe, 8.12–8.13, 8.18
    stainless steel pipe, 8.13–8.14, 8.18
    Streamline, 2.13

K

Kelvin scale, 1.15, 1.16, 1.19, 1.28
Kitchen island, 5.13

L

Labels, 8.6, 8.14–8.15
Laboratories, 4.6, 4.13, 4.14, 8.3, 8.14
Ladder, as a lever, 1.24
Laminar flow, 2.3, 2.17
Lamp, ultraviolet, 3.7, 3.21
Laundry tray
    air gap as backflow preventer, 4.7
    demand and flow pressure, 2.7
    DFU, 6.3, 7.8
    for water softener drainage, 3.12
    WSFU, 2.8

Lavatories
    air gap as backflow preventer, 4.7
    demand and flow pressure, 2.7
    DFU, 6.3, 7.8
    vents for, 5.7, 5.11, 6.11
    WSFU, 2.8
Lawn area, 2.10, 4.9, 4.10, 6.13
Laying out of water supply piping, 2.6–2.10
Lead, 8.8, 8.12
Leader. *See* Conductor
Leaks, 6.6, 7.6
Legionnaire's disease, 4.13
Length
    developed, 2.10, 2.17
    equivalent, 2.3, 2.4, 2.14, 2.17
    units, 1.3, 1.6, 1.7
Lever, 1.23, 1.24, 1.28, 7.3
Liability issues, 4.14
Lift station, 7.16, 7.19
Lime, 3.11, 3.14, 3.21
Limestone, 3.19, 8.15
Liter, 1.10, 1.28
Lock
    air, 7.14, 7.15, 7.19
    trigger, 9.2
Low Hazard classification
    and backflow preventer, 4.5, 4.6, 4.7, 4.14
    definition, 3.22, 4.18
    for pollution, 3.22
Lubricator, 9.9, 9.13

M
Machines
    complex, 1.26
    simple, 1.23–1.26, 1.28
Magnesium, 3.11, 3.14, 3.21
Magnesium salts, 3.2, 3.17, 3.21. *See also* Softening of water
Maintenance and repair
    backflow preventers, 4.14–4.15
    piping system installation for future access, 2.10
    pumps, 7.13–7.15
    sewage and storm water removal systems, 7.13–7.16
    water conditioners, 3.8
Mall, shopping, 6.8
Manganese, 3.14, 3.18, 3.19
Manufactured air gap, 4.11, 4.12, 4.18
Manufacturing plants, 4.11, 8.3
Marble, 3.14, 3.19, 3.21
Material safety data sheets (MSDS), 8.14, 8.15–8.16
Mathematics, applied
    conversions, 1.4–1.5, 1.7, 1.12, 1.30–1.31, 9.4
    expansion coefficient, 1.18
    measuring area and volume, 1.5–1.13
    overview, 1.2, 1.27, 1.28
    simple machines, 1.23–1.26
    temperature, pressure, and force, 1.14–1.23
    weights and measures, 1.2–1.5
Mechanical method of pipe joining, 8.11–8.12, 8.13
Medical and dental facilities, 4.11, 5.3, 7.8
Metals
    heavy, 3.13, 3.16
    in water supply. *See specific metals*
Meter, water, 2.3, 4.10, 4.12
Methane, 5.6
Metric system, 1.2–1.5, 1.16, 1.28, 1.30–1.31
Metrology, 1.2
Microwave ovens, 1.15

Minerals, 3.8, 3.9, 3.11, 3.21
Motors
    air compressor, 9.5, 9.7
    electrical circuit for, 7.15
    pump, 7.12, 7.14, 7.15
    testing, 7.14
MSDS. *See* Material safety data sheets
Mud. *See* Particulates; Turbidity
Municipal facilities, 7.4
Municipal water supply, 3.11, 4.2, 4.7
Municipal water treatment, 3.11
Municipal water utility companies, 3.2, 3.3, 3.11

N
National Association of Home Builders Research Center, 5.8
National Fire Protection Association (NFPA), 7.16, 8.14–8.15, 8.16, 8.19
National Institute of Standards and Technology (NIST), 1.2
National Oceanic and Atmospheric Administration (NOAA), 6.13, 6.14, 7.12
National Physical Laboratory (NPL), 1.2
National Pollution Discharge Elimination System (NPDES), 6.12
National Sanitation Foundation (NSF), 3.8
National Standard Plumbing Code, 2.11
National Weather Service, 7.12
Neoprene, 8.5, 8.9
Neutralizing agent, 8.14, 8.15
NFPA. *See* National Fire Protection Association
NFPA diamond, 8.14–8.15, 8.16, 8.19
NIST. *See* National Institute of Standards and Technology
Nitrates, 3.13, 3.16
Nitric acid, 3.18, 8.5
NOAA. *See* National Oceanic and Atmospheric Administration
Noise
    from air compressors, 9.6
    in the pipes, 1.20, 2.7, 6.6
NPDES. *See* National Pollution Discharge Elimination System
NPL. *See* National Physical Laboratory
NSF. *See* National Sanitation Foundation

O
Occupational Safety and Health Administration (OSHA), 7.15
Odor
    drinking water, 3.2, 3.3, 3.13, 3.14, 3.19, 3.21
    sewage removal system, 7.15
Offset, 5.13, 5.14, 6.5
100-year events, 6.13, 6.14, 6.15, 7.20–7.24
Organic matter, in water supply, 3.19, 3.21
Osmosis, 3.14–3.15, 3.22. *See also* Reverse osmosis systems
Outlet, electrical, 7.16
Overflow
    combined sewer overflow, 6.13, 6.15
    scuppers for, 6.21, 6.24
    strainers, 7.4
Oxidizers, 8.15, 8.16
Oxidizing agent, 3.14, 3.22

P
Parapet around roof, 6.21
Parking lots, 6.13
Particulates, 3.3, 3.9, 3.10, 3.14, 7.4, 7.12
Pasteurization, 3.6, 3.21, 3.22
Patios, 1.19

Paved areas, 6.13
PDI. *See* Plumbing & Drainage Institute
Percolation, 6.22, 6.24
Permeability, 6.22, 6.24
Permit, NPDES, 6.12
Personal protective equipment (PPE)
    compressed air and air-powered tools, 9.2
    sewage pump inspection, 7.6, 7.18
    solvent cement, 8.12
    storm water sump inspection, 7.14, 7.15, 7.18
    testing electric motors, 7.14
    working with hazardous chemicals, 3.5, 8.4
Pesticides, 3.13, 4.2, 4.3, 4.4, 4.10, 8.3
pH, 3.18, 8.2
PHCC. *See* Plumbing-Heating-Cooling Contractors - National Association
Photoelectric cell, 3.7
Photographic processing facilities, 8.3
pi, 1.8–1.9
Pipe
    borosilicate glass, 8.3, 8.4, 8.7–8.9, 8.19
    brass, 3.18, 6.22
    cast-iron, 6.2, 6.21, 6.22, 8.5, 8.12–8.13
    cement or concrete, 6.22
    for compressed air system, 9.8
    copper, 2.13, 3.18, 6.22, 9.8
    corrosive-resistant, 8.2, 8.3–8.6, 8.18
    drainage, 6.5
    DWV. *See* Drain, waste, and vent pipes
    fresh water, 1.10
    glass, 6.22, 8.18. *See also* Pipe, borosilicate glass
    hub-and-spigot, 8.5, 8.12, 8.13
    installation, 8.7–8.14
    movement of fluids through. *See* Flow; Velocity; Viscosity
    no-hub, 8.5, 8.13
    plastic, 1.18, 6.2, 6.22, 8.3–8.5, 8.9–8.12
    polyolefin, 6.22
    Schedule 40 black-iron, 9.8
    silicon, 8.5, 8.12–8.13
    sizing, 2.11–2.14, 6.4
    stainless steel, 1.17, 1.18, 8.6, 8.13–8.14
    steel. *See* Steel
    storage, 3.2
    thermoplastic, 8.4, 8.9, 8.11, 8.18, 8.19
    thermoset, 8.4, 8.5, 8.12, 8.18, 8.19
    underground, 1.18, 6.22
    volume calculations, 1.12–1.13
    water supply. *See* Water supply piping
    wood, 8.5
Pipefitter, 5.8
Piping system. *See* Water supply piping
Piston, 9.6
Pit, holding, 7.2
Pitch, 8.8
Plane, inclined, 1.24, 1.25, 1.28
Plastic
    chlorinated polyvinyl chloride, 7.10, 7.12, 8.4
    couplings, 8.3
    pipe, 1.18, 6.2, 6.22, 8.3–8.5, 8.9–8.12
    polypropylene, 8.4, 8.9
    polyvinyl chloride, 1.18, 6.22, 8.4
    polyvinylidene fluoride, 8.4, 8.5, 8.9
    tetrafluoroethylene, 8.3
    thermoplastic, 8.4, 8.9, 8.11, 8.18, 8.19
    thermoset, 8.4, 8.5, 8.12, 8.18, 8.19
Pliers, as a lever, 1.24

Plugs
    air and mechanical test, 1.22–1.23
    to cap a stub, 6.4
    snap-fit, in air compressors, 9.10
Plumber
    reputation, 3.11
    responsibilities, 1.2, 2.6, 3.2, 3.21, 4.14, 7.9
    water conditioner installer license, 3.14
Plumbing & Drainage Institute (PDI), 1.20, 1.21
Plumbing-Heating-Cooling Contractors - National Association (PHCC), 5.8
Plumbing & Mechanical Magazine, 5.8
PlumbingNet, 5.8
PlumbingWeb, 5.8
Pneumatics, 9.3, 9.13
POE. *See* Point-of-entry unit
Point-of-entry unit (POE), 3.2, 3.22
Point-of-use unit (POU), 3.2, 3.9, 3.16, 3.17, 3.22
Pollution, 3.2, 3.8, 3.9, 3.22, 6.12
Polyester, 8.5
Polyolefin, 6.22
Polyphosphate, 3.18
Polypropylene (PP), 8.4, 8.9
Polystyrene, 8.8
Polyvinyl chloride (PVC), 1.18, 6.22, 8.4
Polyvinylidene fluoride (PVDF), 8.4, 8.5, 8.9
Pond, retention, 7.13, 7.19
Ponding, 6.22, 6.24
Port, atmospheric, 4.14
Potable water. *See* Water supply, potable
POU. *See* Point-of-use unit
Pounds per square inch (psi), 1.19, 1.28
Pounds per square inch gauge (psig), 9.3, 9.12
Power
    backup, 7.12, 7.13
    disconnect prior to power tool maintenance, 9.2
Power failure, 7.14
PP. *See* Polypropylene
PPE. *See* Personal protective equipment
Precipitate, 3.14, 3.21, 3.22
Precipitation (chemical), 3.11, 3.21, 3.22
Precipitation (weather). *See* Rainfall
Precipitation systems, 3.14, 3.21
Pressure
    air compressor, 9.5, 9.8–9.10
    atmospheric (air), 1.19–1.20, 5.3, 5.4, 9.3–9.4
    back. *See also* Backflow preventers; Siphonage, back
        and air admittance vents, 5.12
        conditions which cause, 5.2
        definition, 4.2, 5.17
        in high-rise buildings, 2.7
        and pumps, 7.4, 7.6
    control in compressed air systems, 9.8–9.10
    definition, 1.28
    effect on boiling point, 1.23
    equilibrium in the DWV system, 5.2, 5.3, 9.3
    for fixtures, 2.7
    and force, 1.19–1.23
    in piping system. *See* Water pressure in piping system
    relationship with height, 1.20
    relationship with temperature, 1.19, 9.3
    relationship with volume, 9.4
    subatmospheric, 4.2
    units, 1.3, 9.3, 9.4
    in water heaters, 1.23
Pressure drop, 1.3, 2.12, 2.13, 2.17, 4.7

Pressure-type vacuum breaker (PVB), 4.6, 4.9–4.10, 4.18
Pressure wave, 1.21, 6.4, 9.3
Prism, 1.10–1.11, 1.28, 1.32
Prohibited components, 4.11, 5.8
psi. *See* Pounds per square inch
psig. *See* Pounds per square inch gauge
Public health issues. *See also* Bacteria and viruses
    backflow prevention, 4.2, 4.5, 4.7, 4.14
    clean municipal water supply, 3.2
    corrosive wastes, 8.2
    cross-connection prevention, 4.13
    sewer gas leak, 7.6
Pulley, 1.23, 1.24–125, 1.28
Pulp and paper mills, 8.3
Pumps
    backflow created by failure, 4.2
    backup, standby, or battery-operated, 7.9, 7.12, 7.13, 7.14
    bilge. *See* Pumps, sump
    centrifugal, 7.2–7.4, 7.10, 7.11, 7.13, 7.19
    in chlorinator, 3.3, 3.4, 3.5
    CPVC, 7.10, 7.12
    duplex, 7.4, 7.5, 7.10, 7.19
    history, 7.3, 7.11
    motor, 7.12, 7.14, 7.15
    in pasteurizing system, 3.6
    replacement, 7.15–7.16
    reverse flow, 7.4, 7.5, 7.19
    self-priming, 7.16
    sewage, 7.2–7.6, 7.7, 7.9, 7.19
    storm water, 7.10–7.11. *See also* Pumps, sump
    sump, 6.10–6.11, 6.22, 7.11, 7.19
    water-lubricated, 7.16
PVB. *See* Pressure-type vacuum breaker
PVC. *See* Polyvinyl chloride
PVDF. *See* Polyvinylidene fluoride
Pythagorean theorem, 1.8

R
Rainfall
    conversions, 6.13–6.15
    maximum expected, 6.13, 6.14, 6.15, 6.23
    100-year, 1-hour maps, 6.14, 7.12, 7.20–7.24
Ramps, 1.24
Receiver, air, 9.5, 9.7, 9.13
Receptor, drain, 1.10, 4.11
Reciprocating air compressor, 9.5, 9.6
Rectangle, 1.6, 1.8, 1.10–1.11, 1.13, 1.28, 1.32
Reduced-pressure zone principle backflow preventer assembly (RPZ), 4.11–4.13, 4.18
Refrigeration air dryer, 9.7
Refrigerators, 1.15
Regulators, 2.10, 9.9, 9.13
Reservoirs, 3.11
Residential buildings, 4.13, 6.13, 8.2. *See also* Lawn area
Resin, 3.11, 3.12, 3.13, 8.5
Respirator, 9.2, 9.8
Respiratory disease, 3.5
Restaurants, 4.7
Restrictor, flow, 2.7
Retention strategies
    holding pit from sub-drain, 7.2
    pond, 6.19, 7.13, 7.19
    ponding on the ground, 6.22, 6.24
    roof, 6.19, 6.20
    in a sewage sump, 7.6–7.7
    in a storm water sump, 7.12
    tank, 3.14, 3.15, 7.13, 7.19

Re-vent, 5.10, 5.17
Reverse osmosis systems, 3.14–3.16, 3.19, 3.21
Riser, 8.8, 8.9, 9.10
Roof
    area and rainfall weight calculations, 6.13, 7.12
    pitched *vs.* flat, 5.5, 5.6
    storm drainage systems, 6.13, 6.15–6.22, 6.23, 6.24
    vent through the, 5.2, 5.4, 5.5, 5.6, 6.6
    walls or parapets around, 6.15–6.16, 6.21
Roof plan, 6.18
Roof retention system, 6.19–6.20
Rotary air compressor, 9.5, 9.6
Rounding off numbers, 1.6
RPZ. *See* Reduced-pressure zone principle backflow preventer assembly
Runoff
    agricultural, 4.13, 6.12
    storm water, 6.13–6.22, 7.2, 7.10–7.13, 7.14
Rust, 3.14, 3.21, 8.2, 9.7

S
Safety
    air compressors for breathing apparatus, 9.7
    chlorine, 3.5
    compressed air, 9.2, 9.7
    corrosive wastes, 8.2, 8.4
    electrical, 7.14
    hazard communication, 8.14–8.16
    threaded pipe installation, 9.2
Safety plan, 8.14
Salt
    in water softeners, 3.11, 3.12
    in water supply, 3.19, 3.21
Sand. *See* Particulates; Turbidity
Scale, 3.9, 3.11, 3.17, 8.2
scfm. *See* Standard cubit feet per minute
Scouring, 6.6
Scouring powder, 8.2
Screws
    rotary, 9.5, 9.6, 9.13
    as a simple machine, 1.25, 1.28
    types, 1.25
Scuppers, 6.21, 6.24
Sediment. *See* Particulates; Turbidity
Sensors, 7.9, 7.13, 7.14, 7.15
Septic tank, 1.10, 8.2
Service entrance, 1.19
Sewage. *See* Drain, waste, and vent pipes; Wastewater
Sewage removal systems, 7.2–7.9, 7.15, 7.18, 7.19
Sewer
    combined sewer overflow, 6.13, 6.15
    sizing, 6.5
    storm drain capabilities, 6.13, 7.13
Shaduf, 7.3
Shock absorber, 9.7
Shower
    demand and flow pressure, 2.7
    DFU, 6.3, 7.8
    vents for, 5.11
    water pressure, 2.6
    WSFU, 2.8, 2.9, 2.10
Sidewalks, 1.19
Signage, hazard, 8.14, 8.15, 8.16
Silica gel, 9.7
Silicon, 8.5, 8.12–8.13
Sillcock, 2.7, 3.9, 3.10
Silt. *See* Particulates; Turbidity

Sink
  air gap as backflow preventer, 4.7
  demand and flow pressure, 2.7
  DFU, 6.3, 7.8
  hand, 6.11
  kitchen, 2.8, 6.3, 6.11, 7.8
  service, 2.8
  slop, 1.12
  wash, 6.3, 7.8
  WSFU, 2.8
Siphonage
  back, 2.7, 4.2, 4.3, 4.6, 4.18. *See also* Backflow preventers
  indirect or momentum, 5.2, 5.3, 5.17
  self-, 5.2, 5.3, 5.17
SI system. *See* Metric system
Sizing
  above-grade and below-grade drainage systems, 6.22
  conductors (leaders), 6.16
  drains, 6.4–6.6, 7.7
  DWV system, 6.2–6.11
  history, 2.11
  overview, 2.2, 2.16, 6.2
  pipe calculations, 2.11–2.14, 6.4
  roof drainage systems, 6.15–6.21
  secondary drains, 6.21
  sewage removal systems, 7.9, 7.16
  storm drainage systems, 6.13–6.22, 7.16
  storm water sumps, 7.12
  vents, 6.2, 6.6–6.11
Sleeve, 3.7, 8.8
Slope. *See* Hydraulic gradient
Slug of water, 6.6
Snow, 1.19, 6.6
Soap, 3.17
Soda ash, 3.14, 3.19, 3.21
Soda machines, 4.6, 4.7, 6.11
Sodium, 3.11, 3.13, 3.16, 8.4
Sodium chloride (salt), 3.11, 3.12, 3.19, 3.21
Sodium hypochlorite, 3.3, 3.22
Softening of water, 3.2, 3.9–3.13
Soft water, 3.12, 3.13
Soil
  erosion, 6.12
  permeability, 6.22, 6.24
  stack connected to the, 5.5, 5.6, 5.13, 6.5, 6.8
  and trench depth, 8.9
Solids in wastewater, 7.4
Solvent method of pipe joining, 8.11
Solvents, 3.13, 3.17, 8.3
Sovent® vent systems, 5.14, 5.17
Sprayer, paint, 9.2
Sprinkler system
  fire-suppression, 2.10, 4.6, 4.7
  lawn, 2.10, 4.9, 4.10
Square, 1.6, 1.29
Square foot, 1.6, 1.29
Square meter, 1.6, 1.29
Square root, 1.8, 1.9
Stack
  glass pipe, 8.8
  soil, 5.5, 5.6, 5.13, 6.5, 6.8
  Sovent® vent system innovations, 5.14
  vent (central or main vent)
    definition, 5.4, 5.17, 6.8
    function, 5.4
    location in system, 5.4, 5.5, 5.7
    sizing, 6.4, 6.5, 6.8, 6.11
    vertical water flow mechanics, 6.6
Standard conditions, 9.3, 9.13

Standard cubit feet per minute (scfm), 9.3, 9.13
Steamers, 4.7, 4.10
Steel
  austenitic stainless, 1.17, 1.18, 8.6, 8.18, 8.19
  carbon, 1.17, 1.18
  carbon-molybdenum, 1.17, 1.18
  duplex stainless, 8.6, 8.18, 8.19
  effects of corrosive wastes, 8.2
  ferritic stainless, 8.6, 8.18, 8.19
  galvanized, 6.22, 9.8
  intermediate alloy, 1.17
  low-chrome, 1.17, 1.18
  martensitic stainless, 8.6, 8.18, 8.19
  stainless, 8.3, 8.5, 8.6, 8.18
  thermal expansion, 1.17, 1.18
Sterilizer, 4.13
Stills, 3.16–3.17
Storm drainage systems, 6.13–6.22
Storm water, 6.13–6.22, 7.2, 7.10
Storm water removal systems, 7.2, 7.10–7.13, 7.15–7.16, 7.18, 7.19
Strainers
  in backflow preventer assembly, 4.11, 4.15
  installation for future access, 2.10
  overflow, 7.4
  roof drain, 6.20
  in sump for wastewater solids, 7.4
Streets, 6.13
Stroke, 3.3, 3.22
Stub, branch, 6.4
Sub-drain, 7.2, 7.12, 7.16, 7.19
Sulfates, 3.17
Sulfides, 3.14
Sulfur, 3.3, 3.14, 3.19, 3.21
Sulfuric acid, 3.18, 8.4, 8.5
Sump
  acid dilution and neutralization, 3.19, 8.14, 8.15, 8.18
  definition, 7.19
  sewage, 7.6–7.7, 7.16, 7.18. *See also* Pumps, sump
  storm water, 7.12–7.13, 7.16
Supports, 1.18, 8.13, 9.8
Swimming pool, 1.10, 1.12, 3.3, 4.9
Switches
  float, 7.7, 7.9, 7.13, 7.14, 7.19
  pressure, 9.7
  probe, 7.9, 7.19
  time delay, 3.7
Système International d'Unités (SI). *See* Metric system

T
Takeoff, 2.6, 9.10
Tanks
  air storage, 9.6
  flush, 2.7, 2.8, 2.9, 2.11
  neutralizing, 3.19, 8.14, 8.15
  processing, 4.13
  retention, 3.14, 3.15, 7.13, 7.19
  settling, 3.11, 3.19
  water closet, 2.7, 2.8
  water storage, 1.22
Tannins, 3.9
Tape, heat, 4.7
Taste, foul, 3.2, 3.3, 3.13, 3.14, 3.19, 3.21
Tee
  90-degree, friction loss due to, 2.4
  sanitary, 6.9
  test. *See* Cleanout
  threaded, 8.11
Teflon®, 8.5

Temperature
    conduction of heat energy, 1.14
    definition and use of term, 1.2, 1.14, 1.29
    effects on water supply piping, 2.2–2.3
    measurement, 1.15–1.16
    pasteurization, 3.6
    relationship with pressure, 1.19, 9.3
    scales of measurement, 1.15–1.16, 1.19, 1.28
    in water heaters, 1.23
Terminal, vent, 5.4, 5.5, 5.17
Test cock, 4.10, 4.15, 4.18
Tests
    air, for glass pipe, 8.7
    backflow preventers, 4.14–4.15, 4.17
    DWV system, 1.22–1.23
    individual pressure zones, 4.10, 4.15
    for iron and manganese, 3.18
    pH, 3.19
    turbidity, 3.19
    water hardness, 3.17–3.18
    water supply, 3.3, 3.11
Tetrafluoroethylene plastic (TFE), 8.3
TFE. *See* Tetrafluoroethylene plastic
ThePlumber.com, 5.8
Thermal expansion and contraction, 1.15, 1.16–1.18, 1.29, 8.3, 8.12
Thermometers, 1.15–1.16, 1.28, 1.29
Thermoplastic, 8.4, 8.9, 8.11, 8.18, 8.19
Thermoset, 8.4, 8.5, 8.12, 8.18, 8.19
Thermostat, 2.3, 3.6, 7.9
Thread, 8.11–8.12, 9.2
Throttle, 9.9, 9.13
Toilet, 1.24, 3.18, 5.6, 5.12
Toluene, 3.17
Tools
    English *vs.* metric system, 1.4
    pneumatic, 9.2, 9.3
Tower, cooling, 4.9
Training
    backflow preventer tester, 4.15
    medical gas venting design, 5.3
    safety with hazardous chemicals, 8.14, 8.16, 8.18
    water conditioner installer, 3.14
Transient flow, 2.3, 2.17
Traps
    grease, 1.10, 1.12, 6.11
    from storm drain to sewer line, 6.13
Trap seal, 5.2, 5.12
Trench, 8.9
Triangles
    isosceles, 1.8, 1.28
    Pythagorean theorem, 1.8
    right, 1.8, 1.11, 1.28, 1.32
Troubleshooting, 3.17–3.19, 7.13–7.16, 7.18
Turbidity, 3.9, 3.11, 3.16, 3.19, 3.21
Turbidity unit, 3.19, 3.22
Turbulent flow, 2.3, 2.6 , 2.17

U

ULHS. *See* Uniform Laboratory Hazard Signage
Ultraviolet light, 3.7–3.8, 3.21, 3.22
Underground systems, 1.18, 6.22, 8.8–8.9, 8.13
Uniform Laboratory Hazard Signage (ULHS), 8.15, 8.19
Uniform Plumbing Code™ (UPC), 5.12, 7.16
University of California Foundation for Cross-Connection Control and Hydraulic Research, 4.3
Unloading, 9.9, 9.13
UPC. *See Uniform Plumbing Code*™

Urinals
    demand and flow pressure, 2.7
    DFU, 6.3, 7.8
    vents for, 6.11
    water pressure, 2.6
    WSFU, 2.8
U.S. Department of Agriculture (USDA), 7.12
U.S. Environmental Protection Agency (EPA), 4.4, 6.12, 6.15, 8.2
U.S. Food and Drug Administration (FDA), 3.3
USDA. *See* U.S. Department of Agriculture
Use factor, 9.10, 9.13
Utility companies. *See* Water utility companies

V

Vacuum, partial, 5.2, 5.4, 7.4
Vacuum breaker
    atmospheric, 4.6, 4.8–4.9, 4.18
    hose connection, 4.6, 4.18
    in-line, 4.14, 4.18
    pressure-type, 4.6, 4.9–4.10, 4.18
    spill-proof, 4.6
Valves
    air admittance, 5.12–5.13, 5.17
    air receiver, 9.5, 9.6
    air throttle, 9.9, 9.10, 9.13
    automatic drain, 9.10
    backwater, 7.13
    blowdown, 9.10
    bypass in water softener, 3.12, 3.13
    check
        in backflow preventers, 4.2, 4.10–4.11, 4.12, 4.13, 4.14
        in pumps, 7.2, 7.4, 7.5, 7.10, 7.13, 7.15, 7.16
        in sump, 7.6, 7.7, 7.16
    in chlorinators, 3.4, 3.5
    cutoff, 7.4, 7.5
    diaphragm exhaust, 7.14
    float-controlled, 1.24
    flush, 2.8, 2.9, 2.11
    in flushometer, 2.7
    friction loss due to, 2.3, 2.4
    gate, 4.10, 7.6, 7.7, 7.10, 7.13, 7.16
    handwheel on, 1.26
    inlet and outlet in water softener, 3.12, 3.13
    installation for future access, 2.10
    mixing, 2.8
    pressure-reducing, 2.7
    in pumps, 7.2, 7.10, 7.13
    relief, 4.11, 4.12, 9.5, 9.6
    shutoff, 3.9, 4.10, 4.11, 9.9
    solenoid, 3.6, 3.7, 3.22
    in sumps, 7.6
    temperature/pressure relief, 1.23, 2.2–2.3
    test cock, 4.10, 4.15, 4.18
Vehicle maintenance facilities, 8.3
Velocity, 2.3
Venting systems
    combination waste and vent system, 5.13, 5.17, 6.11
    for hazardous chemicals, 8.14
    reduced size, 6.11
Vents
    air admittance, 5.12–5.13, 5.17
    back, 5.10, 5.17
    battery, 5.11, 6.10
    branch, 5.10, 5.17
    circuit, 5.5, 5.11, 5.17, 6.10
    common (unit), 5.5, 5.10, 5.17, 6.8, 6.9
    continuous, 5.10, 5.17

crown, prohibited, 5.8
definition, 5.17
extension to prevent frost, 6.6
flat, 5.4, 5.17
grade or slope, 5.5–5.7
history, 2.11
how they work, 5.2
individual, 5.10, 5.17, 6.8
intermediate atmospheric, 4.6, 4.13, 4.18
loop, 5.11, 5.13, 5.17
main or central (vent stack). *See* Stack, vent
pneumatic ejector relief, 5.14
purpose, 1.20, 5.2, 5.16, 6.2
relief, 5.5, 5.13, 5.14, 5.17, 6.10
roof, 5.2, 5.4, 5.5
rule of thumb, 5.16
sewage sump, 7.6
sizing, 6.2, 6.6–6.11
Sovent® vent systems, 5.14, 5.17
stack (highest horizontal), 5.4, 5.5, 5.13, 5.17
sump, 6.10–6.11
system design, 5.2–5.8
types, 5.10–5.14
wet, 5.11–5.12, 5.17, 6.8–6.10
yoke, 5.13, 5.17, 6.10
Vibration, 1.20
Vinyl ester, 8.5
Viscosity, 2.3, 2.17
VOC. *See* Volatile organic compounds
Volatile organic compounds (VOC), 3.17
Voltage supply, 7.14
Volume
air conversions, 9.4
calculations for containers, 1.12–1.13, 7.12
definition, 1.29
liquid, 1.3, 1.10
measuring, 1.10–1.13, 1.32
and pipe diameter, 1.12
rainfall conversions, 6.13, 6.15
relationship with force, 1.20
relationship with pressure, 9.4
units, 1.3, 1.11, 1.12
of water, effects of temperature, 2.2

W
Walls, 6.15–6.16, 6.21, 8.8
Washing machine
DFU, 6.3, 6.8, 7.8
vents for, 6.11
WSFU, 2.8
Waste stack, 5.2
Wastewater. *See also* Drain, waste, and vent pipes
aeration in Sovent® vent system, 5.14
in combined sewer overflow, 6.15
contamination of potable water. *See* Backflow; Backflow preventers; Siphonage, back
corrosive, 8.2–8.3
hazardous. *See* Hazardous materials
holding areas. *See* Retention strategies
sewage removal systems, 7.2–7.9, 7.15
sizing sewers, 6.5
sizing storm water sumps, 7.12
solids in, 7.4
temperature considerations, 1.14
treatment plants, 4.11
weight considerations, 1.12
Water
and the Celsius scale, 1.16
density, 2.2–2.3
discoloration, 3.18
foul taste, 3.2, 3.3, 3.13, 3.14
hard, 3.9, 3.11–3.13, 3.16, 3.17–3.18, 3.21
moisture in air lines, 9.7, 9.10, 9.12. *See also* Dryer, air
odor, 3.2, 3.3, 3.13, 3.14
ph. *See* Acids and acidity; Alkalines and alkalinity
physical properties, 2.2–2.3, 2.16
potable. *See* Water supply, potable
scalding, 8.2
soft, 3.12, 3.13
weight at specific temperatures, 2.2
weight of specific volumes, 1.12
Water closet
backflow preventers, 4.6, 4.10
demand and flow pressure, 2.7
DFU, 6.3, 6.4, 6.5, 7.8
pump or discharge pipe, 7.16
vents for, 5.11
WSFU, 2.8, 2.10
Water drip leg, 9.10, 9.12, 9.13
Water hammer, 1.20, 1.21, 9.3
Water heaters
damage by hard water, 3.9, 3.11
explosion, 2.2
isolation from ion exchange system, 3.12
temperature and pressure, 1.23
thermometers used in, 1.15, 1.16
thermostatic probe, 7.9
Water pressure in piping system
booster systems, 2.6
excess over friction loss, 2.12, 2.14
hammer, 1.20, 1.21, 9.3
international standards, 2.6
loss due to friction, 2.3–2.4
minimum and maximum, 2.12
Water quality
contamination. *See* Contamination
pollution, 3.2, 3.8, 3.9, 3.22, 6.12
public health issues, 3.2, 4.2, 4.5, 4.13
storm water runoff hazards, 6.13, 7.10, 7.14
Water Quality Association (WQA), 3.8
Water-saving measures, 1.12, 2.7
Water seal. *See* Trap seal
Water service pipes, 1.18
Water supply, potable
bottled, 3.3, 4.13
codes and standards, 2.2
contamination with wastewater. *See* Backflow; Backflow preventers; Siphonage, back
disinfection, 3.2–3.8, 3.11, 3.21
filtration and softening, 3.9–3.17
municipal, 3.11, 4.2, 4.7
sterilized water, 3.2, 4.13
treatment, 3.2, 3.11
troubleshooting, 3.17–3.19
from wells, 3.2, 3.11, 3.15, 4.10
Water supply fixture unit (WSFU), 1.20, 1.21, 1.29, 2.8–2.9
Water supply piping
air in the, 4.2
damage by hard water, 3.9
factors affecting, 2.2–2.4
filter locations, 3.9, 3.10
flushing, 3.2–3.3, 3.10
hot and cold, 2.11
laying out, 2.6–2.10
roughing-in, 3.10
sizing, 1.12, 2.11–2.14, 2.16
Water tower, 1.22
Water treatment facility, 3.11

Water utility companies, 3.2, 3.3, 3.11
Weather considerations, 5.5. *See also* Freezing conditions; Rainfall
Websites for information, 5.8, 6.13, 8.15
Wedge, 1.25, 1.29
Weight
   in calculation of liquid volumes, 1.12
      rainfall, 6.13, 6.15
      units, 1.3
Weir, trap, 5.6
Welding, 8.13
Well water, 3.2, 3.11, 3.15, 4.10
Wheel
   of a pulley, 1.24, 1.25
   as a simple machine, 1.23, 1.26, 1.29
   water, 7.3
Wheelbarrows, 1.23, 1.24, 1.26
Whirlpool, 2.10, 4.10
Wood, 8.5
Wool, glass, 8.8
Work, 1.23, 1.29
WQA. *See* Water Quality Association
Wrench, 1.24, 1.25, 8.7
WSFU. *See* Water supply fixture unit
Wye, 8.11

Z
Zeolite, 3.11, 3.12–3.13, 3.14, 3.18, 3.21